STRENGTH OF THE DIAMOND-METAL INTERFACE AND BRAZING OF DIAMONDS

STRENGTH OF THE DIAMOND-METAL INTERFACE AND BRAZING OF DIAMONDS

Yu V Naidich, V P Umanskii and I A Lavrinenko

I.N. Frantsevich Institute of Materials Science
National Academy of Sciences of Ukraine, Kiev

Viva Books

New Delhi | Mumbai | Chennai | Kolkata | Bangalore | Hyderabad | Kochi

First Indian Edition 2008

VIVA BOOKS PRIVATE LIMITED

- 4737/23 Ansari Road, Daryaganj, New Delhi 110 002
  E-mail: vivadelhi@vivagroupindia.net, Tel. 42242200
- Plot No. 76, Service Industries, Shirvane, Sector 1, Nerul, Navi Mumbai 400 706
  E-mail: vivamumbai@vivagroupindia.net, Tel. 27721273, 27721274
- Jamals Fazal Chambers, 26 Greams Road, Chennai 600 006
  E-mail: vivachennai@vivagroupindia.net, Tel. 28290304, 28294241
- B-103, Jindal Towers, 21/1A/3 Darga Road, Kolkata 700 017
  E-mail: vivakolkata@vivagroupindia.net, Tel. 22836381, 22816713
- 7, Sovereign Park Apartments, 56-58, K. R. Road, Basavanagudi, Bangalore 560 004
  E-mail: vivabangalore@vivagroupindia.net, Tel. 26607409, 26607410
- 101-102 Mughal Marc Apartments, 3-4-637 to 641, Narayanguda, Hyderabad 500 029
  E-mail: vivahyderabad@vivagroupindia.net, Tel. 27564481, 27564482
- Beevi Towers, First Floor, SRM Road, Kaloor, Kochi 682 018, Kerala
  E-mail: vivakochi@vivagroupindia.net, Tel: 0484-2403055, 2403056

www.vivagroupindia.com

Published by arrangement with
Cambridge International Science Publishing
7, Meadow Walk, Great Abington
Cambridge CB1 6AZ, UK

ISBN-10: 81-309-0838-7
ISBN-13: 978-81-309-0838-0

Published by Vinod Vasishtha for Viva Books Private Limited, 4737/23, Ansari Road, New Delhi-110 002. Printed and bound by Anand Sons, Delhi-110 092.

Contents

Chapter 1

PHYSICAL–CHEMICAL PROCESSES OF JOINING DISSIMILAR MATERIALS

1.1 GENERAL ADHESION PROBLEMS

The relationships governing the mechanism of bonding of dissimilar materials are very complicated. The experimental results, including the results of experiments carried out at the I.N. Frantsevich Institute of Materials Science, National Academy of Sciences of Ukraine, have been used to develop general principles and theoretical prerequisites interpreting the processes of adhesion and bonding dissimilar materials.

Bonding in adhesion contact in various cases is explained by molecular (van-der-Walls), electrostatic and chemical interaction, and also by mechanical fixing where one of the components is implanted in surface irregularities of the other component. The bonding strength is often the result of the effect of several of these factors. The effect of each factor changes depending on the nature of materials, their properties and technological parameters of the bond.

To realise some interaction, it is necessary to ensure efficient contact of phases at the contact surfaces — bring the surfaces to the distance of action of atomic binding forces, 10^{-10}–10^{-9} m. Interfacial contact in the solid–solid system can be ensured at:

1) contact of flat surfaces of two solid phases (materials);

2) contact of the surafces of the solid and liquid with subsequent solidification of the latter (crystallisation, transition to the hard glassy state, polymerisation, drying, etc);

3) deposition from the gas phase of one or several substances on the solid surface of another phase..

If the contacting phases (materials) are represented by solid non-plastic substances, the true contact area even in the case of a smooth flat surface of both phases represents a negligible part of the total interfacial area and there is almost no bonding between the phases (clean and dry surfaces are considered).

The true contact area rapidly increases if one of the phases is liquid (case 2), or can be plastically deformed. Consequently, the atoms of the liquid phase, filling the micro- and macroirregularities of the solid component, form perfect contact at the interface. This is also observed in case 3.

In the conditions of perfect contact between dissimilar solids the latter interact on molecular and atomic levels.

Van-der-Walls forces, combining polarisation, dipole, or dispersion interaction act between the electrically neutral systems (atoms, molecules, crystals) when there is no electron exchange or hybridisation of the electron orbitals of the dissimilar atoms. This interaction is universal and acts in all systems. Van-der-Walls interaction is determined by relatively long-range forces (10^{-9} m). The binding energy of these forces is not high — fractions or units of kilojoules per mole. The expression for the binding energy of the dispersion forces between the pair of atoms 1 and 2, situated at distance R has the form

$$E \cong \frac{3}{2}\frac{\alpha_1\alpha_2}{R^6}\cdot\frac{I_1 I_2}{I_1 + I_2}, \qquad (1.1)$$

where α and I are the polarisability and the first ionisation potential of interacting atoms.

Neglecting the entropy term and taking into account only the interaction of the nearest atoms of the first atomic planes of the contacting phases, it is possible to estimate the binding energy of two substrates (phases): $W_A \cong nE$, where n is the number of atoms of the substance situated on the unit surface area.

The role of the image potential interaction in adhesion that can act between the metal (electroconductive media) and electrocharged atoms is debatable and for diamond (nonpolar substrate)/metal system is probably not significant.

The dispersion interaction can determine the specific (in some cases sufficient for practice) mechanical strength of the contact. The strongest contact is ensured by the effect of chemical forces (homeo- or heteropolar interaction). The binding energy of these forces is an order of magnitude higher than the binding energy of the van-der-Walls forces and equals tens and hundreds of kilojoules per mole (kJ/mole). These forces are determined by the exchange or hybridisation of the electrons of the interacting atoms and act over a distance of approximately 10^{-10} m. To generate this interaction, the atoms must be brought together to this distance. For chemical forces this is more difficult to achieve than for the van-der-Walls interaction which acts over longer distances. Therefore, the negative effect of films and contaminants of the contact surfaces in the case of chemical interaction is generally stronger than for systems with physical interaction.

The problem of the effect of films separating the interlayers on the adhesion of the contacting surfaces is of special-interest and will be examined below. In some cases the binding energy of the contacting solid can be calculated. Theoretical prerequisites for this (examined in detailed in Ref. 6) are as follows.

The contact systems are subdivided into equilibrium and non-equilibrium. In the former, the contacting phases (solids) are under the learning equilibrium conditions, i.e. the chemical potential of the components in each phase are

$$\mu_i^{I} = \mu_i^{II}.$$

In these systems, there is no interaction between the contacting phases (dissolution, formation of new chemical compounds).

In the non-equilibrium contacting systems, where the chemical potentials of the components in different phases are not equal,

$$\mu_i^{I} \neq \mu_i^{II},$$

interfacial chemical interaction takes place – dissolution, mutual diffusion, formation of interlayers of new phases (compounds).

In the general case, the work of adhesion (binding energy) between the contacting phases is

$$W_A = W_{A\,eq} + W_{A\,noneq}. \tag{1.2}$$

The contribution $W_{A\,eq.}$ consists of the energy of van-der-Walls (dispersion) interaction and the equilibrium chemical part of the binding energy determined by the closure of free valencies of the atoms on the surfaces of the contacting phases; this contribution is significant for interatomic bonds of the contacting solids of similar nature (in particular, for intermetallic systems) and is almost insignificant for metal-non-metallic solid and, in particular, metal–diamond crystal systems.

Thus, the binding energy between the phases, especially the metal–non-metallic solid, of greatly differing nature is

$$W_A = W_{A\,dis} + W_{A\,chem.reac.} \tag{1.3}$$

Both these terms can be evaluated theoretically. The value $W_{A\ dis}$ is calculated using equation (1.1). To calculate $W_{A\ chem.noneq.}$, it is necessary to know the energy and the equation of the interfacial reactions taking place. These values are approximately identical with the energy generated during a chemical reaction between the atoms of the first atomic planes of the contacting phases, and can be calculated on the basis of a thermodynamic approach.

If the atomic plane of phase A with surface density n_A mole is in contact with the atomic plane of the phase B containing n_B mole of substance per 10^{-4} m^2, and if the reaction

$$mA + nB \leftrightarrow A_m B_n, \qquad (1.4)$$

takes place in the A–B system, the total work of reaction W_A, i.e. the binding energy, is

$$W_A = \Delta z = \int_0^{\alpha_0} dz, \qquad (1.5)$$

where $dz = f(C_i, \alpha)d\alpha$ is the expression for the isotherm of the reaction as a function of the concentration of the reacting substances C_i and the degree of transformation of the reactions. α is the amount of the compound formed, mole.[6] After integration, we obtain the expression for W_A as a function of α_0 of the equilibrium degree of transformation. The last quantity is determined from the expression for the equilibrium constant of the reaction

$$\Delta z^0 = -RT \ln k = -RT \ln \frac{C^0_{A_m B_n}}{C_A^{0m} C_B^{0n}}, \qquad (1.6)$$

where Δz^0 is the standard change of the isobaric potential of the reaction.

Thus, the following characteristics of the system must be available to calculate $W_{A\ noneq.}$: n_A, n_B, the approximate reaction mechanism must be known and the change of the standard isobaric potential during this reaction Δz^0 must be available. Examples of calculating the values of W_A for the case of physical and chemical interaction in the metal–carbon systems are given below.

If one of the phases is liquid, the binding energy can be experimentally determined by investigating wettability. The contact wetting angle θ and the surface tension of the liquid phase in the system σ_{liq} are determined. The results are then used to calculate the work of adhesion between the phases:

$$W_A = \sigma_{liq}(1 + \cos\theta) \tag{1.7}$$

For practice, however, it is important to know the force characteristic of the strength of the interface — the mechanical strength of the bond. The ratio of the force and energy characteristics of the system is examined in chapter 2.

1.2 INTERACTION AND WETTING IN THE DIAMOND–METAL SYSTEM

In interaction of carbon with metal, the dependence on the properties of the metal shows different types of bonds between carbon and metal atoms.[12-16]

The electronic structure of the carbon atoms ($1s^2 2s^2 2p^2$ – four valency electrons) is such that, depending on conditions, carbon can be either an electron donor or acceptor.

The metalloid properties of the carbon atoms are usually manifested in interaction of carbon with strong electropositive elements, i.e. elements situated in the left part of the periodic system of elements, – alkaline and alkaline–earth metals with a low ionisation potential. These elements form salt-like compounds with carbon – carbides with a high fraction of ionic bond between the Me–C atoms. For example, alkaline metals, implanted into the graphite lattice in the space between the layers, undergo positive ionisation, forming polycarbides of the type, MeC_8, MeC_{16}.[13,17-20]

According to Ref.13, *d*-electrons of the metal are transferred on the bond between the carbon atoms, and with decreasing ionisation potential of the metal atom carbon atoms form complicated sublattices anions. The dimensions of the atoms (ions) of the alkaline metals are also important.

According to the data in Ref.18,19, the energy of formation of the carbide of the alkaline metal consists of the energy of separation of graphite layers in implantation of the metal ion and of the energy of the metalc–carbon bond. The first quantity brings a negative contribution to the total energy owing to the fact that energy must be used up for increasing the distance between the graphite hexagons. Therefore, as the radius of the metal ion decreases the separation energy become lower and the energy of carbide formation increases.

The alkaline–earth metals also form with carbon compounds of the type of salt-like carbides MeC_2 with the ionic component being the main type of bond between the metal and carbon atoms.

In interaction of carbon with transition metals, metals with the unfilled *d*-electron shell (*d*-metals), carbon shows its metallic donor properties, transferring part of its external electrons to the *d*-band of the metal. This assumption, initially formulated by Ubbelode[21] for hydrides of transition metals, was developed later in Ref. 14, 15, 22, 23 for the interaction of carbon with a transition metal in carbide phases. The transition of part of valency electrons of the carbon atoms to the *d*-band of the metal – positive ionisation of carbon in bonding with the *d*-metal – is supported by many facts. They include the metallic nature of the properties of the carbide phases (interstitial phases), the small dimensions of the carbon ions and their high mobility in diffusion in the carbides (this mobility is considerably higher than that of the anions of oxygen or sulphur in oxides or sulphides), the relationship of the heat of formation of the carbides and the degree of filling of the *d*-band of the metal, the large difference of the actually observed lattice parameters of the titanium and vanadium carbides in comparison with those assumed on the basis of calculations for the Ti^+– C^- and V^+–C^- ionic bond.[15]

According to Bruer,[24] the fact that carbon is an electron donor supplying electrons to the *d*-band of the metal is confirmed by the observed displacement of the maximum of stability (atomisation heat) of solid phases which for the transition metals is localised in groups V and VI of the periodic system (formation of stable electronic structures d^5), and for the carbides in group IV.

Covalent bonds with carbon are formed mainly by the metallic elements containing external *p*-electrons in the energy levels with the main quantum numbers 2 and 3. The characteristic representatives of this group are boron and silicon. These elements form relatively stable carbides.

As the main quantum number increases to $n \geq 4$, the strength of the *p*-metal–carbon covalent bond decreases so rapidly that these elements are almost completely inert in relation to carbon.

According to the nature of interaction with carbon, the metals with incomplete *f*-electron shells and also actinoids are placed in a separate group.[13] These elements form salt-like covalent-metallic carbides. The low values of the first ionisation potential of the lanthanides result in a high fraction of ionic bonds of metal with carbon. On the other hand, these metals have unfilled *f*-shells. Like in the case of the *d*-transition metals, this should make the bond to be of the metallic type;

indeed, this is manifested in the carbides of this type. The latter have the most typical properties of the metals – high electrical conductivity of the metallic type.

The actinoids form even more metallic bonds with carbon. After all, all elements without exception interact with carbon due to physical forces – dispersion interaction. The binding energy, determined by these forces is, as already mentioned, low and, consequently, the role of these forces is significant only in systems in which the level of the carbon-metal chemical interaction is relatively low. These elements include the metals of auxiliary B-subgroups of the periodic system of the periods IV–VI:

Group I	Group II	Group III	Group IV	Group V	Group VI
Cu	Zn	Ga	Ge	As	Se
Ag	Cd	In	Sn	Sb	Te
Au	Hg	Tl	Pb	Bi	Po

Some of these elements, for example, silver, copper, gold can, under certain conditions, form salt-like carbides, but these compounds are highly unstable and often dissociate with an explosion.[12,16]

Depending on the intensity of interaction, i.e. the level of binding forces of the metal and carbon, and also on some other factors, especially the dimension factor, the metals should also be classified on the basis of the macrocharacter of their interaction with graphite and diamond. The following main types of contact interaction can be defined:

I. The metal–diamond (graphite) interface is characterised by the formation of an interlayer of a new carbide phase with a relatively small change of the composition, at least, after a short contact time. This type of interaction includes the transition metals of the groups IV – VI, and also silicon, boron and, evidently, some alkaline–earth metals.

II. Dissolution of the substance of the solid phase (diamond, graphite) in liquid metal: the metal does not dissolve and do not diffuse into the solid phase. This type of interaction is found in the metals of groups VIII and, to some extent, VII of the periodic system.

III. The diffusion and implantation of the metal into the lattice of the solid phase (graphite). This reaction is found in systems formed by graphite with alkaline metals. Experimental results show that boron can also diffuse into graphite.[25] No interaction of this type has been observed for diamond.

IV. The absence of any chemical interaction, constant interface. It is evident that in accordance with the considerations made regarding this group, metals chemicall inactive in relation to carbon belong in this group (B–subgroup of the periods IV–VI of the Periodic table of elements).

Adhesion interaction of liquid metals with the surface of diamond and graphite

Taking into account the above-described classification of the interaction of metal and carbon on the basis of the types of binding forces and the macrocharacter of the interfacial reaction, we shall examine the experimental data obtained for the wetting and contact interaction in the diamond (graphite) – metallic melt, grouping the metallic elements of the basis of the identical nature of their contact properties. Since the binding forces between the carbon atoms in the graphite lattice in the layer are not lower that those in the diamond lattice, as indicated by the low transition heat of the diamond–graphite modification (2.1 kJ/mole) when a sublimation heat is approximately 711 kJ/mole, in the energy sense the two modifications are similar and the energy effect of chemical reactions in them is also similar. Therefore, in the interaction with liquid metals, the diamond and graphite differ only slightly, regardless of the large difference in their structure, as confirmed in Ref. 5 and 26. This makes it possible to obtain additional information of the wetting capacity of diamond using similar data for graphite.

Metals of auxiliary B-subgroups of the periodic system of elements of the periods IV–VI

These elements are almost inert in relation to carbon. They do not form stable carbides.[12,16] In the liquid state up to the boiling point, they usually dissolve small amounts of carbon (10^{-3}–10^{-1}),[27–30] and do not attack graphite materials during long-term contact.[31,32] Of these metals, the wetting behaviour in relation to diamond has been investigated for copper, silver, gold, gallium, indium, germanium, tin, antimony, and bismuth. These metals do not wet diamond (and graphite), the contact wetting angles are 120–150°. The results of examination of the wetting of various crystallographic faces (111) and (100) of diamond (copper, silver, gold, germanium and tin) showed that the metals wet the (111) face slightly better.[33] The binding energy (the work of adhesion) of the liquid metal–carbon surface is low (70–300 mJ/m^2, 2–6 kJ/mol), Table 1. Experiments show a constant (gallium–graphite, indium–diamond, tin–diamond systems) or even a slight reduction (copper–graphite system) of the work of adhesion with temperature (Fig. 1, 2).[6] If it is assumed that this interaction is of the chemical nature, then owing to the fact that the process is obviously endothermic, it should be expected that the

Table 1 Molar work of adhesion of liquid metals to diamond

Liquid metal	t, °C	W_{M} kJ/mol
Cu	1100	4.2
Ag	1100	7.9
Au	1100	2.1
In	1000	1.7
Ge	1200	6.7
Sn	1150	3.3
Pb	400	4.6
Sb	900	2.9

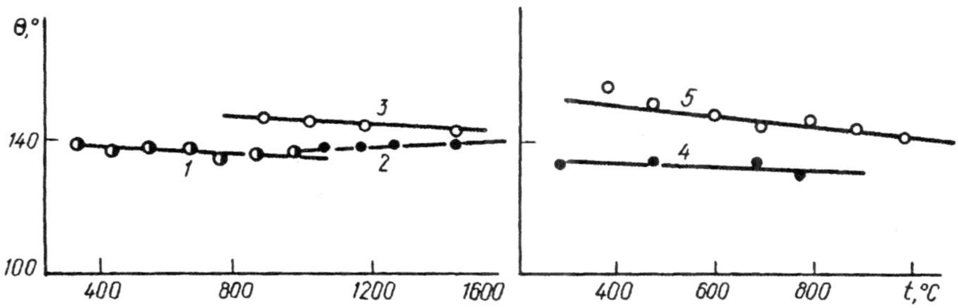

Fig.1 Temperature dependences of the contact wetting angle in systems: 1) indium-diamond; 2) copper–graphite; 3) tin–graphite; 4) tin-diamond; 5) gallium–graphite.

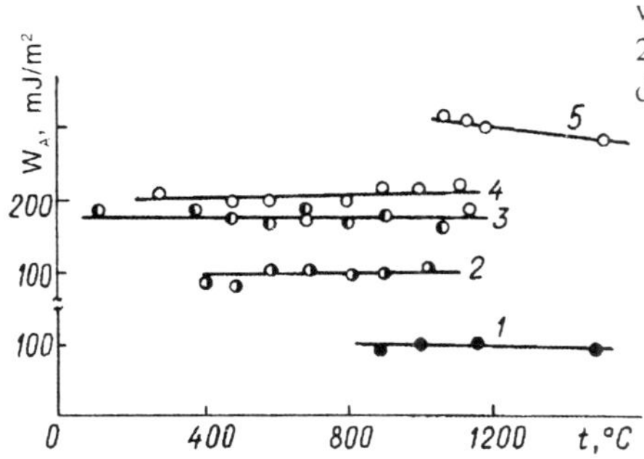

Fig.2 Temperature dependences of the work of adhesion in systems: 1) tin–diamond; 2) indium–diamond; 3) gallium–graphite; 4) tin–graphite; 5) copper–graphite.

intensity of interaction (the work of adhesion) will increase with temperature. However, this has not been observed. These facts can be explained assuming that the physical interaction is controlling for such systems (graphite, diamond–metal of the auxiliary B-subgroup).

The examined systems with slight interaction in the examined temperature range are characterised by a thermodynamically equilibrium contact of almost pure components – the metal and carbon. This means that the value of $W_{A\ noneq.}$ is low.

Using the equation for the potential of the dispersion forces, it is possible to estimate quantitatively the work of adhesion for this case. The quantity R was assumed to be equal to approximately $R_{Me} + R_C$, where R_{Me} and R_C is the half of the minimum atomic spacings in the metal and graphite. Polarisability was determined approximately from the relationship

$$\alpha \cong \frac{e^2h^2}{4^2ml^2}. \qquad (1.8)$$

The interaction energy per unit of the interfacial surface is

$$E \cong eN. \qquad (1.9)$$

(where N is the number of metal–carbon bonds per unit of the surface).

Since the structure of diamond contained $0.345 \cdot 10^{20}$ of carbon atoms per 10^{-4} m^2, and for the metals examined here this value varies in the range $0.080–0.170 \cdot 10^{20}$ atom/m^2, then to a first approximation the number of the bonds established at the liquid metal–carbon surface contact is determined by the number of metal atoms.

As indicated by the data in Table 2, there is a certain correspondence between the calculated energies of interaction at the solid–liquid interface and the work of adhesion W_A. In any case, dispersion interaction is sufficient to ensure the experimentally observed adhesion energy. It should be noted that according to the classification of the macrocharacteristic types of contact reactions, described previously, this is the fourth type of interaction.

Elements chemically interacting with carbon

The metals, chemically interacting with carbon, are characterised by high binding energy with the diamond surface. The transition carbide-forming elements are preferred in this case because the interfacial reaction in these systems results in the formation of a carbide interlayer at the interface. This interlayer is characterised by the metallic atomic bond and metallic properties. This is beneficial for producing strong bonds with metals.

Table 2 Comparison of calculated values of energy E and experimental determined values W_A

Metal	$R_{Me} \times 10^{10}$, m	$I \times 10^{19}$, J	$\alpha \times 10^{30}$, m^3	$e \times 10^{20}$, J	E, mJ/m²	W_A at 1000°C, mJ/m²	
						Diamond	Graphite
Cu	1.28	12.3	1.81	2.73	480	230	315
Ge	1.39	10.4	2.54	2.14	320	390	100
Ga	1.39	9.6	2.98	2.80	420	–	160
Ag	1.44	12.2	1.86	1.65	230	450	250
In	1.57	9.3	3.19	1.79	210	100	105
Au	1.44	14.7	1.27	1.33	185	120	–
Sn	1.58	11.7	2.02	1.21	140	190	70
Sb	1.61	13.6	1.49	0.95	105	180	80
Pb	1.75	11.8	1.96	0.81	75	265	75
Bi	1.82	11.5	2.07	0.72	65	–	95

Transition metals

A number of transition metals form strong compounds – carbides – with carbon. Less active of these metals dissolve large amounts of carbon in the liquid state.

The elements Ti, Cr, V, Mn, Nb, W, Mo, Fe, Co, Ni, Pd, Pt, Rh, Ta, Re, Hf[5,6,34–37] Zr,[38] U,[39] Ce[40] examined so far in the pure form or in the form of small additions to non-active elements show high adhesion activity in contact with diamond and graphite.

The work of adhesion of these metals with respect to the surface of diamond and graphite is high and reaches 2500–3000 mJ/m². Increase of temperature greatly increases in the adhesion of diamond and graphite to such metals.

The adhesion activity of these metals in relation to diamond (and graphite) differs. The comparison of these values for different metals by examining their wetting behaviour in the pure form is very difficult because of differences (sometimes very high for a number of elements) in the melting point, and also due to the fact that many transition metals in the pure form completely spread over the surface of graphite (and diamond) and the work of adhesion cannot be determined numerically.

In these cases it is necessary to examine the solutions of these metals in a solvent inert in relation to carbon. The concentration dependence of adhesion and wetting of alloys is determined in relation to the diamond surface.

The solvent is represented by copper, tin, gallium, germanium, gold and other metals. The solvent has often the form of several inert metals (for example, copper–tin, copper–silver, copper–gallium, etc).

Titanium is characterised by very high adhesion activity. A small addition of titanium to alloys with tin, copper, in copper–tin–titanium and copper–silver–titanium ternary alloys greatly increases the work of adhesion (Fig.3).[6]

Chromium also acts highly efficiently in alloys with copper,[41] gallium,[42] and in copper–gallium alloys.[43]

The authors of Ref. 44 examined the wetting and work of adhesion of a number of such metals and alloys in relation to diamond. Using the Ag–28.0% Cu alloys as an example, it was shown that the copper–silver alloys do not wet diamond (θ = 130°) and the work of adhesion of these alloys is low (400 mJ/m^2). At the same time, the addition of titanium to the alloys resulted in a large decrease of the contact wetting angles and an increase of the work of adhesion (Fig.4).

In the tin–titanium system, the liquid phase for tin-rich alloys forms at a temperature of 200 °C, but wetting of diamond is observed at 800 °C and higher temperatures. Figure 5 shows the concentration dependences of the contact wetting angle of the diamond and graphite by the tin–titanium alloys. As indicated by the results, already a small addition of titanium (0.4%) decreases the contact wetting angle on diamond to 20–30° in comparison with 125° for pure tin.[34]

The investigators found[45] that an addition of titanium to a copper–tin alloy, which does not wet the diamond (θ = 130–140°), in an amount of 5 to 20%, greatly decreases in the contact wetting angle (Fig. 6) and increases the work of adhesion to 2000 mJ/m^2.

Similar data were obtained in Ref.1,5,46,47. As reported, pure copper does not wet the diamond surface. However, after adding more than 0.1% of Cr to copper, 11.7% titanium or 7.7% V the resultant binary

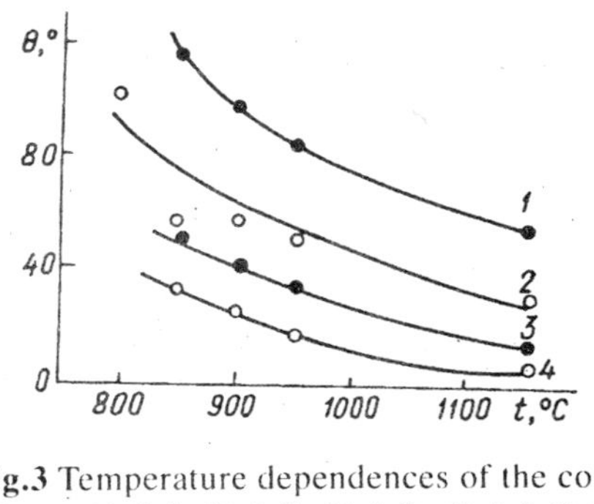

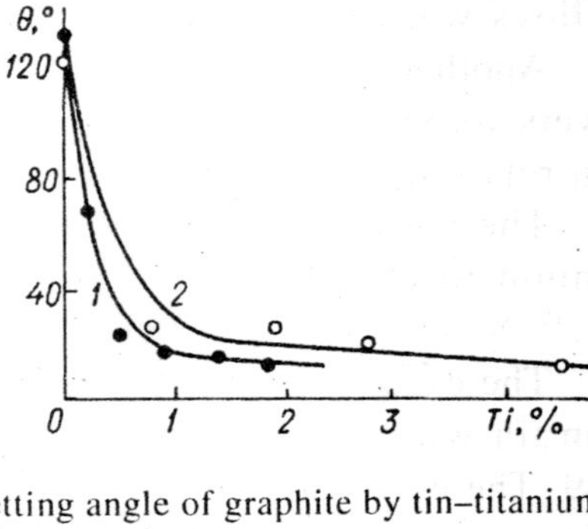

Fig.3 Temperature dependences of the contact wetting angle of graphite by tin–titanium alloys: 1) 0.5; 2) 1.0; 3) 1.9; 4) 2.9 % Ti.

Fig.4 Contact wetting angle of diamond (1) and graphite (2) by the Cu–72% Ag alloy with titanium additions.

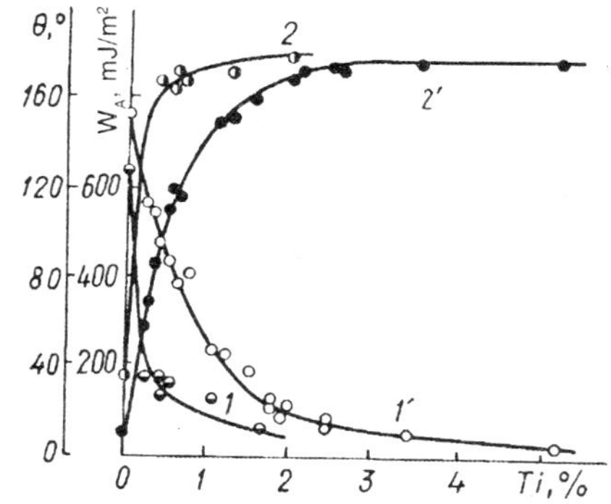

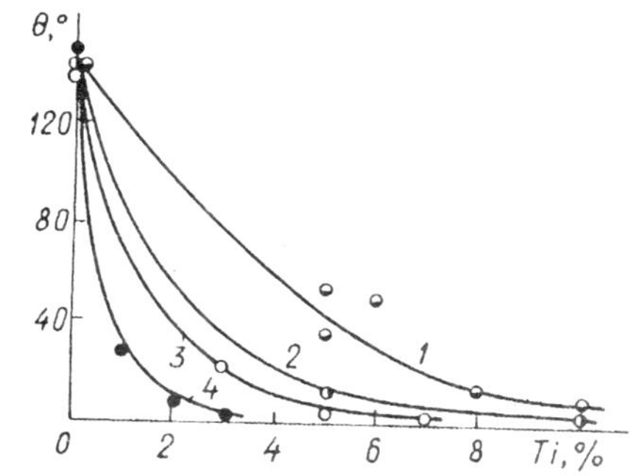

Fig.5 Concentration dependence of the contact wetting angle of tin–titanium alloys on diamond (1), graphite (1') and the work of adhesion (2, 2') at 1150 °C.

Fig.6 Concentration dependence of the contact wetting angle of diamond by Cu–Sn–Ti alloys at 1150 °C (τ = 20 min) and the tin content of: 1) 0; 2) 10; 3) 20; 4) 25%.

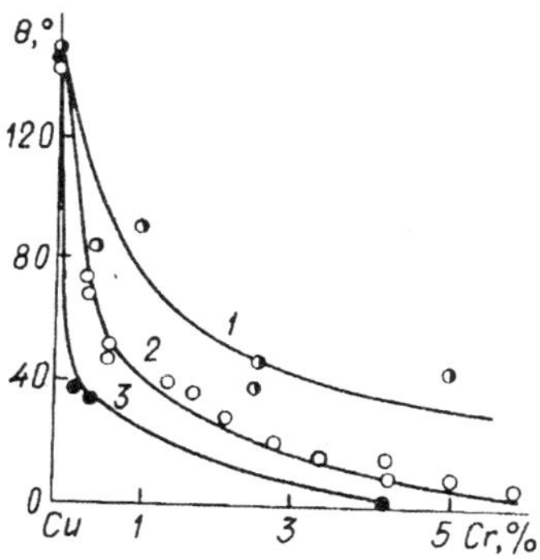

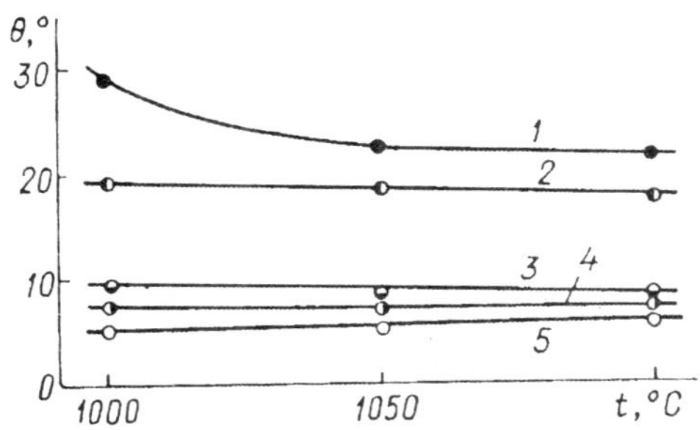

Fig.7 Concentration dependences of the contact wetting angle of graphite and diamond by copper–chromium alloys: 1,2) graphite at 1150 (1) and 1250°C (2); 3) diamond at 1150°C.

Fig.8 Wetting of graphite by Cu–18.8% G–Cr melts: 1) 0.4; 2) 0.6; 3) 0.8; 4) 1.0; 5) 1.2% Cr.

alloys wet the diamond sheet, Fig. 7.[7]

Another alloy inactive in relation to diamond, Cu–18%Ga, also poorly wets its surface (θ = 151–158°). The work of adhesion of the alloy in relation to diamond is only 90–150 mJ/m^2.

The addition of even smaller amounts (0.8%) of titanium or chromium to the alloy increases the wetting and work of adhesion, Fig. 8.[47,48]

The examination of the kinetics of wetting of graphite by alloys based on tin with additions of zinc and titanium has been described in Ref. 49. The experimental results show that the additions of titanium improve the wetting of graphite, Fig. 9. This also relates to antimony-based alloys, Fig. 10. Titanium, absorbed from the melt on the graphite surface, reduces the interfacial energy in the contact zone thus decreasing the

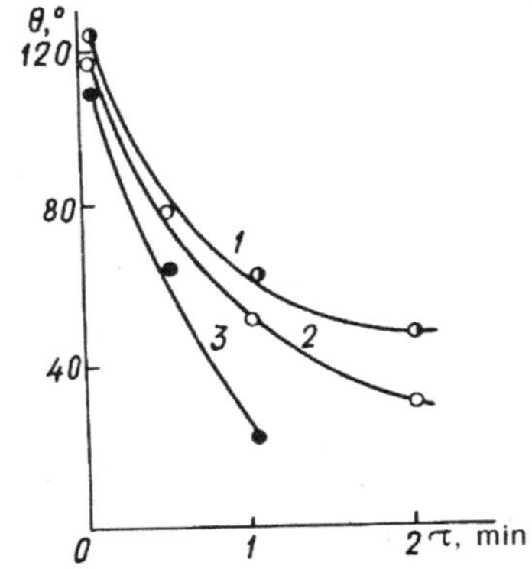

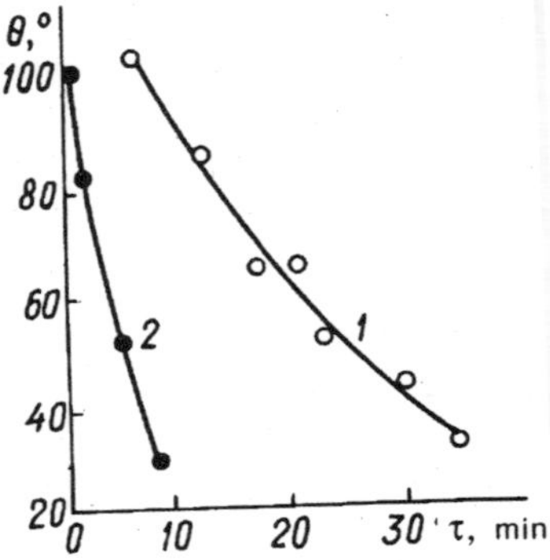

Fig.9 Kinetics of wetting AG-1500 graphite in helium by Sn – 3.0% Zn – 3.0% Ti melt: 1) 970; 2) 1020; 3) 1070 K.

Fig.10 Kinetics of wetting AG-1500 graphite in vacuum by Sb – 1.0% Zn – 5.0% Ti melt: 1) 1070; 2) 1120 K.

contact wetting angle.

In Ref. 5–9, 50 it was shown that other transition metals – zirconium, hafnium, vanadium, niobium, tantalum, molybdenum, tungsten, manganese – also support the interaction of the alloys, in which they are included, with the surface of graphite and diamond. The product of interaction in this case is the carbide phase distributed at the boundary of diamond with the metal or alloy.

The results of microscopic examination of the carbide phase, in particular in the Cu–Cr system, where this phase can be easily found (Fig.11) show that the phase grows towards the side of the liquid alloy from the contact surface and its microhardness is 14200–15000

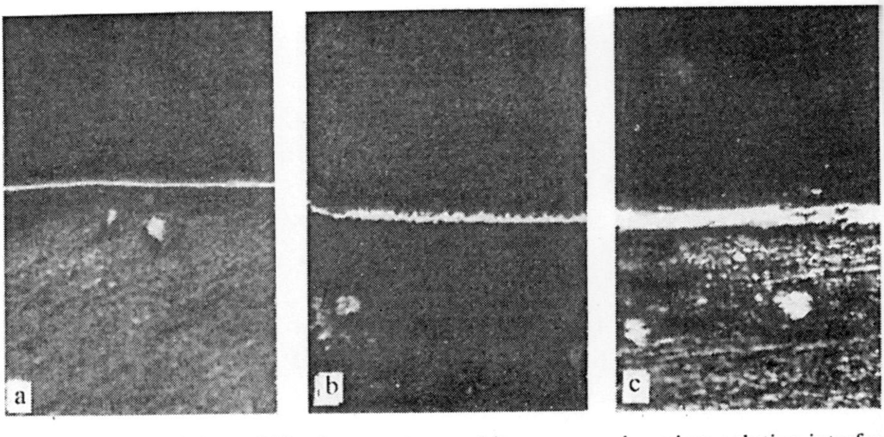

Fig.11 Interlayer of the carbide phase at the graphite–copper–chromium solution interface at different temperatures and contact times (×500): a) 5; b,c) 25 min (a,b – 1150; c – 1250°C); upper parts are metal.

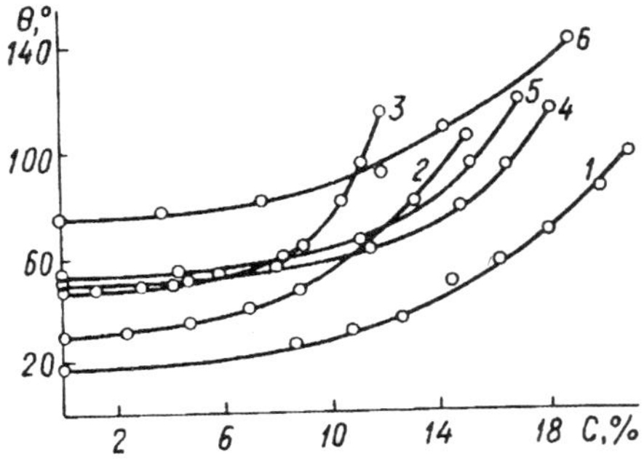

Fig.12 Concentration dependences of the contact wetting angle of graphite by melts of: 1) Fe-C; 2) Co-C; 3) Ni-C; 4) Rh-C; 5) Pd-C; 6) Pt-C.

MPa, which is close to the hardness of Cr carbides, and x-ray diffraction analysis shows that this phase consists of chromium carbides.[51]

Far less attention has been given to the wetting of systems consisting of diamond and metals of the groups VII or VIII of the periodic table of elements.[6,52] The results of examination of wetting of graphite by iron, rhodium, palladium, and platinum, carried out in Ref.6 and 36 show that the pure metals efficiently wet graphite as a result of the reaction of dissolution of carbon in the metal. Preliminary saturation of these metals with carbon greatly decreases the wetting capacity. This is in agreement with the corresponding data[53,54] for nickel and cobalt. The values of the work of adhesion for pure and carbon-saturated metals of group VIII decrease with increasing order number of the element (Fig.12).

In Ref. 55, in examining the wetting of graphite and diamond by alloys containing metals with different chemical activity for the solid phase (Ni,Mn,Cr,V,Nb,Ta,Ti) it was established that the adhesion activity of the metals-additions increases in the series from nickel to titanium in accordance with the increase of the chemical affinity of metals for carbon (Fig. 13).

In Ref. 56–58, the authors examined in detail the wetting and contact interaction in the graphite–Ni–Ti, Ni–Cr and Ni–V–melts systems. As expected, the addition of carbide-forming elements to nickel increases the wetting of graphite by the metallic melt. However, it is interesting to note that the C–Ni–Cr system does not show preferential buildup of the carbides in the zone of contact with graphite, whereas in the C–Ni–V system at approximately 20 at.% of vanadium a continuous layer of vanadium carbide forms at the interface. The thickness of this layer increases with increasing vanadium concentration of the melts. As regards the graphite–Ni–Ti system, high wetting is observed here when the titanium content of nickel reaches 10 at.%, and then up to approximately 65 at.% of titanium wetting does not change. It then starts to increase, reaching complete wetting ($\theta = 0^o$) at approximately

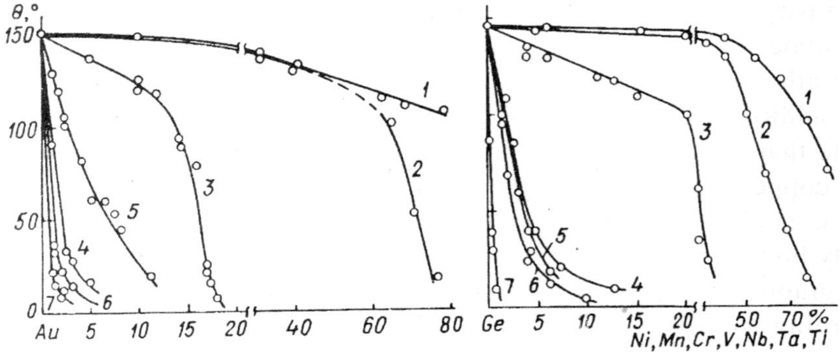

a b

Fig.13 Concentration dependences of the wetting of graphite by melts of gold (a) and germanium (b) at 1200°C with additions of transition metals: 1) Ni; 2) Mn; 3) Cr; 4) V; 5) Nb; 6) Ta; 7) Ti.

80 at.% of titanium in the melt. The authors link this phenomenon with the equilibrium diagram of the Ni–Ti system characterised by the existence of intermetallic compounds with a high enthalpy of formation. This leads to strong negative deviations from the ideal thermodynamic activity of titanium in the melts and impairs the wetting of graphite. In the region of the equilibrium diagram in which intermetallic phases do not form, graphite is complete wetted by the Ni–Ti melts.

Investigations[59] into the effect of the thermodynamic activity in alloys, especially manganese in alloys with gold, tin, copper and silver, on the wetting of diamond show (Fig.14) that there is a specific relationship between the thermodynamic and adhesion activity of the melt component. The highest degree of wetting should be expected in systems combining high chemical affinity of the component for the solid phase with its high thermodynamic activity in the melt (low binding energy between the metal and the solvent). In this connection, it is interesting to consider the results obtained in Ref.60. The authors examined the wetting of graphite by Cu–Mn melts with additions differing in the

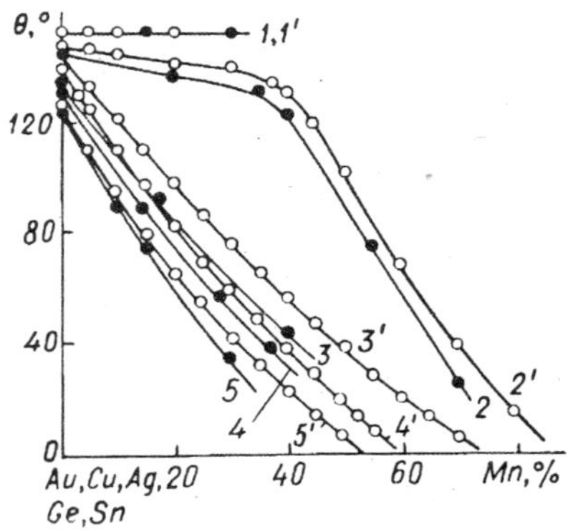

Fig.14 Wetting of diamond (1 – 5) and graphite (1' – 5') by melts (1100 °C): 1, 1') Au – Mn; 2, 2') Ge – Mn; 3, 3') Sn – Mn; 4, 4') Cu – Mn; 5, 5') Ag – Mn.

intensity and nature of the interaction with the components of the base – copper or manganese. The results show that the addition of lead to the alloy has almost no effect on the wetting of graphite, whereas a tin addition greatly reduces wetting (Figs. 15, 16). It is interesting to note that the simultaneous presence of lead and tin in the alloys based on copper and manganese increases the interfacial activity of manganese at the boundary with graphite and leads to the results considerably higher than the overall results. Differences in the nature of wetting of graphite by the Cu–Sn–Mn and Cu–Pb–Mn melts are explained, according to the authors, by differences in the thermodynamic activity of manganese in these alloys which is considerably lower in the Cu–Sn alloys in comparison with Cu–Pb alloys, due to the presence of intermetallic compounds in the Sn–Mn system and the absence of such compounds in the Pb–Mn system. Similar investigations, carried out previously in Ref. 61 for Cu–Sn–Ti alloys, show that the increase of the tin content of the alloy greatly increases the interfacial activity of titanium. This also relates to the Cu–Ti alloys with additions of silver and gallium.[62,63]

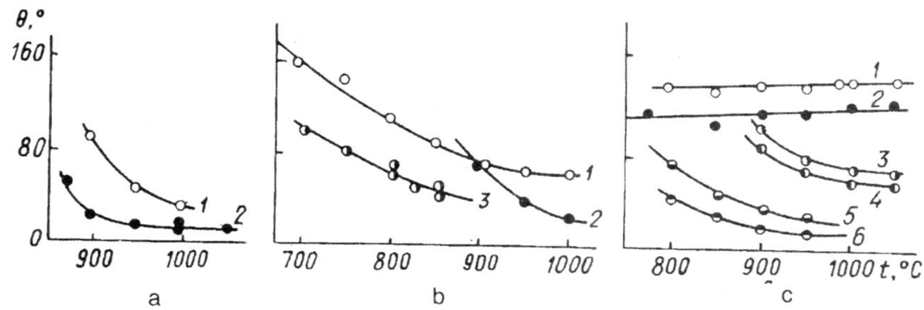

Fig.15 Temperature dependences of the contact wetting angle of graphite by melts: a – 1) 80Cu–20Mn; 2) 60Cu–40Mn; b – 1) (80Cu–20Sn) – 2) 75Cu–25Mn; 3) (80Cu–20Sn)–24Mn–6Ti; c) – 1 (80Cu–20 Sn)–10Mn, 2– (80Cu–20Sn)–20 Mn; 4) (80Cu–20Pb)–20Mn; 5) (80Cu–10Sn–10Pb)–20Mn; 6) (80Cu–10Sn–10Pb)–20Mn–4Ti.

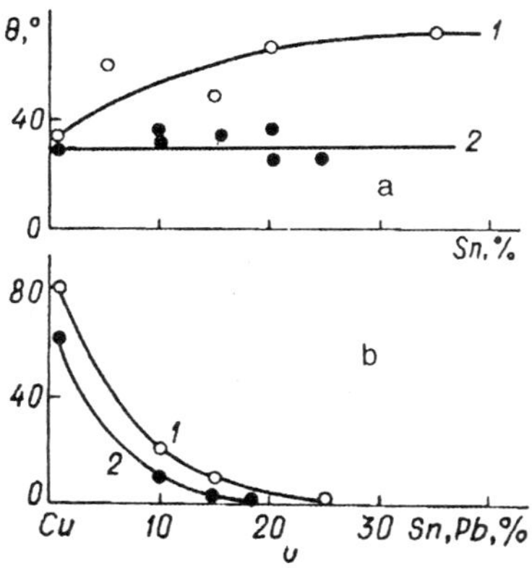

Fig.16 Concentration dependences of the contact wetting angle of diamond at 1100°C (a) and graphite at 950°C (b) on the tin or lead content of ternary alloys: a – 1) (Cu–Sn) – 6Ti; 2) – (Cu–Sn) – 7Ti; b – 1) (Cu–Sn) – 25Mn; 2) (Cu–Pb) – 25Mn.

The authors of Ref. 64 examined the effect of adding an inactive component to Ni–Ti alloys on the wetting of graphite. This element is capable of greatly reducing the solubility of carbon in the melt thus influencing the nature of interaction in the interfacial region. At a constant Ni:Ti ratio of 95:5, they change the germanium concentration in the melt from 0 to 100%. The result show that intensive wetting of graphite by the Ni–Ti–Ge melt was observed at 30–50% of germanium, corresponding to 3.5–2.5 at.% of titanium in the melt. The results indicate that the thermodynamic activity of titanium in the ternary alloy is higher as a result of intensification of the bonds between other components of the alloy by nickel and germanium and due to corresponding weakening of their bonds with titanium.

Thus, the addition of a third inactive element to the binary adhesion-active alloys may exert differing effects on the behaviour of these alloys in contact with the solid phase: the third component may prove to be neutral and can increase or decrease the interfacial activity of the adhesion-active component.

Analysis and generalisation of the results of examination of wetting and contact interaction in the carbon material–metallic melt systems, characterised by the occurrence of chemical processes at the interface, have been carried out in Ref. 65. The author has formulated a number of assumptions on physical–chemical relationships governing high-temperature capillarity and the nature of driving forces of the processes taking place in these systems and determined methods of changing the wetting and adhesion at the contact of between the carbon material and the melt.

When examining the wetting of graphite and diamond by multicomponent alloys[66] containing two adhesion-active components, for example manganese and titanium in Cu–Sn–Mn–Ti and Cu–Sn–Pb–Mn–Ti melts, or even three components Mn, Ti and Ni in the Ni–Mn–Sn–Ti alloy,[67] the results show improvement of wetting in comparison with similar alloys not containing titanium (Fig.17). The effect of adhesion-active additions in the examined alloys is added up.

The experimental results show that the adhesion activity of transition metals can be compared with the thermodynamic potential of the carbide formation reaction or carbon dissolution in the melt.

On the basis of decreasing adhesion activity of transition metals in relation to diamond, the metals can be arranged in the following series: Ti, Cr, V, Mn, Fe, Co, Ni for period IV: for period V Zr, Nb, Mo, Rh, Pd; for period VI W, Pt.

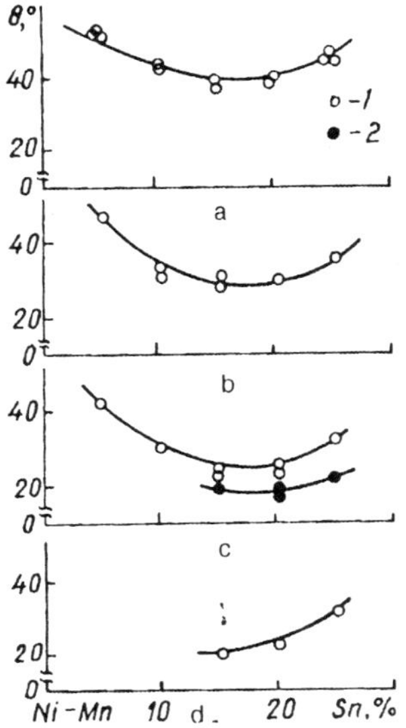

Fig.17 Dependence of the contact wetting angle of graphite (1) and diamond (2) on the tin content of the Ni–Mn–Ti alloy and a titanium content of: a) 0; b) 5; c) 10; d) 15%.

Metals forming covalent bonds with carbon

This group includes silicon, aluminium, boron and also beryllium. The first three metals show quite high adhesion activity in relation to the surface of graphite and diamond; the adhesion work reaches approximately 1000–1200 mJ/m^2. However, these values are lower than the maximum values of the binding energy with graphite or diamond observed for the transition metals of, for example, period IV (approximately 2000–3000 mJ/m^2) which corresponds to a lower heat of the reaction $p\text{Me} + q\text{C} \rightarrow Me_pC_q$ for these metals in comparison with transition metals (for example, 52 for SiC, 190 kJ/mol for TiC).

The spreading of these metals on the surface of graphite or diamond is caused by the chemical reaction of carbide formation at the interface (reaction type I).

In accordance with the general theoretical examination of the wetting of solids by metallic liquids, it is possible to calculate the adhesion work of chemically active metals in relation to the surface of graphite and diamond. We shall use equation (1.3):

$$W_A = W_{A\,VdW} + W_{A.chem.noneq.} \tag{1.10}$$

where $W_{A\,VdW}$ is the energy of the van-der-Walls interaction of the metal and graphite (diamond) which can be calculated using the same procedure as for metals inert in relation to carbon.

For the case of systems with carbide formation at the interface, the equation of the reaction of metal Me and graphite C can be written in the form

$$pMe + qC \leftrightarrow Me_pC_g. \tag{1.11}$$

If the initial amounts (in moles) of the metal and carbon (the number of moles per unit contact surface) are respectively n_{Me} and n_C, the amount of each substance at any moment of the reaction is $n_{Me} - p\alpha$ for the metal, $n_C - q\alpha$ for carbon, α for the compound, where α is the degree of transformation of the number of moles of the formed carbide, i.e. the work of adhesion for the given system, can be calculated using the following equations:

$$W_A = \int_0^{\alpha_0} \left[\begin{array}{l} RT \ln \dfrac{\alpha(n_{Me} + n_C - \alpha V)^V}{(n_{Me} - p\alpha)^p (n_C - q\alpha)^q} - \\ -RT \ln \dfrac{\alpha_0(n_{Me} + n_C - \alpha_0 V)^V}{(n_{Me} - p\alpha_0)^p (n_C - q\alpha_0)^q} \end{array} \right] d\alpha. \tag{1.12}$$

Here α_0 is the equilibrium degree of transformation:

Calculating the integral, we obtain

$$W_A = -RT\left(n_{Me} \ln \frac{1 - \dfrac{p}{n_{Me}}\alpha_0}{1 - \dfrac{V}{n_{Me} - n_C}\alpha_0} + n_C \ln \frac{1 - \dfrac{q}{n_C}\alpha_0}{1 - \dfrac{V}{n_{Me} + n_C}\alpha_0} \right). \tag{1.13}$$

The value α_0 is determined from the equation for the standard isobaric potential of the reaction

$$\Delta z = -RT \ln k = -RT \ln \frac{\alpha_0(n_{Me} + n_C - \alpha_0 V)^V}{(n_{Me} - p\alpha_0)^p (n_C - q\alpha_0)^q}. \tag{1.14}$$

Results of calculating the adhesion work for several simple systems are presented in Table 3.

Similar calculations can also be carried out for systems where the boundary reaction, which determines the wetting in the system, is the dissolution of carbon in the metal. For this purpose, it is necessary to have data on the variation of the free energy in the formation of the carbon–metal solution.[6]

Metals forming ionic bonds with carbon

This group includes alkali, alkali-earth and also rare-earth metals. The quantitative data on the wetting of graphite by the metals of this group are available for lithium at 300 °C (θ = 80°). It is also well-known that at temperatures of about 450 °C sodium forms a wetting angle smaller than 90° on graphite. Alkali metals in long-term contact with graphite greatly erode its surface.

Table 3 Theoretical calculations of the work of adhesion and the contact wetting angle in the metal – graphite system (ionisation potential of carbon 17.9 × 10^{-19} J, number of cabon moles per 10^{-4} m^2 of the area 0.39 × 10^{-4}

System	T, K	n_{Me} × 10^{12} mol/m^2	I_{Me} × 10^{19}, J	R × 10^{-10}, m	ΔZ, kJ	$W_{A(VDW)}$, mJ/m^2	$W_{A(chem)}$, mJ/m^2	$W_{A(VDW)}$ + W_A(chem), mJ/m^2	σ, mJ/m^2	θ, ° Calculated	θ, ° Experimental
Ti–C	2000	0.231	10.9	2.15	–155	230	3270	3500	1460	0 Positive spreading coefficient	0 Positive spreading coefficient
Zr – C	2200	0.19	11.0	2.34	–159	120	2850	2970	1390	As above	As above
Si – C	1725	0.23	13.0	1.89	–42	400	830	1230	720	45	20 – 30*
Al – C**	1500	0.223	9.6	2.14	–44	280	1100	1380	800	44	39

* Initial value of the contact angle at holding τ → 0.

** In calculations instead of the equation of the reaction 4/3Al + C → $Al_{4/3}C$ it was assumed approximately that Al+C ⇄ AlC.

The contact of graphite with alkali metals is accompanied by the formation of polycarbides of metals of the MeC_8, MeC_{16} type with the negative heat of reaction, although its absolutely value is small. For example, for the reaction of graphite with liquid potassium

$$C_{sol} + K_{liq} \rightarrow KC_4$$

ΔH = –6276 kilojoules/mol of carbon.

However, the equation of the reaction can be written only approximately and the value of ΔH is highly inaccurate (according to other data, for the same reaction ΔH = –3766 J/mol); in addition, the values of ΔH for other systems have not been determined quantitatively.

Thus, the numerical calculations of the adhesion work and the wetting in such systems are not possible at the moment. Nevertheless, taking in the account the low values of the surface tension of alkali metals (100–400 mJ/m^2) and the negative values of the heat of reaction, it can be assumed that these systems should be characterised by the extent wetting (contact angle lower than 90°) which is also observed in lithium and sodium.

The graphite-alkali metal systems shows the third type of interaction: diffusion of the metal into graphite.

The literature contains no data on the wetting properties of alkali-earth metals in relation to graphite or diamond. The high negative values of the heats of formation of carbides of the corresponding metals (–120.1 kJ/mol for BeC_2; –44.0 for MgC_2; –29.3 for CaC_2; –25.1 for BaC_2) should lead to wetting in the system. An exception may be magnesium whose carbide is unstable.

On the basis of these considerations it is possible to carry out a qualitative and, in some cases, quantitative prediction of adhesion activity in relation to the surface of graphite or diamond by the metals for which the experimental data are not available (Fig. 18).

It's interesting to count the number of metallic elements chemically active in relation to carbon and wetting or not wetting the surface of diamond (graphite). Of approximately 88 metals of the periodic system of elements, 18 elements, i.e. approximately 20% show very low, almost nonexistent chemical activity for carbon and, consequently, the very low adhesion activity and the absence of wetting of the surface of graphite and diamond (θ = 130–140°). These metals are inert in relation to carbon. The remaining 70 metals show different degrees of chemical and, correspondingly, capillary activity sufficient for a high degree of wetting of the surface of graphite and diamond. It should be noted that the number of the metals which do not interact with carbon is relatively high. In this respect, carbon occupies, as regards in-

H										He
⊕ Li	⊕ Be	⊕ B	C	N	O	F				Ne
⊕ Na	⊕ Mg	⊕ Al	⊖ Si	P	S	Cl				Ar
⊕ K	⊕ Ca	⊕ Sc	Ti ⊕	V ⊕	Cr ⊕	Mn ⊕	Fe ⊕	Co ⊕	Ni ⊕	
Cu ⊖	Zn ⊖	Ga ⊖	⊖ Ge	⊖ As	Se	Br				Kr
⊕ Rb	⊕ Sr	⊕ Y	Zr ⊕	Nb ⊕	Mo ⊕	Tc ⊕	Ru ⊕	Rh ⊕	Pd ⊕	
Ag ⊖	Cd ⊖	In ⊖	⊖ Sn	⊖ Sb	⊖ Te	I				Xe
⊕ Cs	⊕ Ba	⊕ La	Hf ⊕	Ta ⊕	W ⊕	Re ⊕	Os ⊕	Ir ⊕	Pt ⊕	
Au ⊖	Hg ⊖	Tl ⊖	⊖ Pb	⊖ Bi	Po	At				Rn
⊕ Fr	⊕ Ra	⊕ Ac	Th ⊕	Pa ⊕	U ⊕					

Fig.18 Periodic system of elements showing the degree of adhesion activity of metals in relation to diamond and graphite.[6] The plus sign shows the strong adhesion activity of the metal (contact angle lower than 90°); the minus sign indicates weak adhesion activity (contact angle higher than 90°). The frame indicates the metals examined in experiments.

ertness, one of the leading positions amongst all other elements, evidently, after noble gases and hydrogen.[6]

Since diamond is a metastable modification at usual pressures, it is of considerable interest to examine its wetting by metals and adhesion under the conditions of the thermodynamic stability of diamond. These investigations have been carried out in Ref. 68. Wetting was examined at pressures of 5–8 GPa to 2500 °C. High-pressure apparatus was used: metal droplets were formed in the a sodium chloride melt which is inert in relation to diamond and metals and transfers the pressure. This investigation was carried out on single crystals of natural diamond (group VIII) with a size of approximately 1 carat. The wetting of the natural face (111) was investigated. The angles were measured on the basis of the shape of the solidified droplet. The results of special investigations show that the error in determining the true contact wetting angle is not high: it does not exceed 10%.

The metals investigated were copper, silver, gold which do not wet diamond under the vacuum conditions, nickel, palladium, platinum which dissolve carbon in an amount of 0.10–0.16 molar fractions and wet diamond.

The chemical inertness of the medium (the sodium chloride melt) causes that the shape of the metal droplet and the value of the contact angle are determined mainly by the interaction of the metal with the diamond surface. The resultant values of the contact wetting angles do not differ greatly from the wetting angles of diamond by the metal in the vacuum or an inert gas. The data for the diamond single and polycrystals are presented in Table 4. The value of adhesion

Table 4 Wetting of diamond polycrystals by metallic melts at superhigh pressures

Metal	P, GPa	t, °C	θ, °	Metal	P, GPa	t, °C	θ, °
Ni	7.0	1740	45	Cu	5.5	1300	150
Co	7.0	1720	28	Au	5.8	1400	135
Fe	7.0	1740	26	Ag	5.8	1400	140
Pd	8.0	2000	64	Ni	8.0	2500	39
Pt	8.0	2100	65	Pd	8.0	2500	46
Rh	8.7	2300	44	Pt	8.0	2500	58

in the systems is high for transition metals (2000–3000 mJ/m^2) and low for copper, gold and silver (200–400 mJ/m^2).

It should be noted that with increasing temperature the wetting of diamond by non-active metals increases. For example, the contact wetting angle of copper at temperatures above 2870 K decreases to 90° which is evidently associated with the increase of the solubility of carbon in copper at such high temperatures and pressures (Fig.19).

The results obtained in this work are of considerable interest for determining the work of formation of a diamond nucleus, its critical dimensions, the nucleation rate of diamond, i.e. the parameters which that must be known for diamond synthesis.

The investigations of the nature and structure of interphase layers and contact interaction of diamond (graphite) with metal melts (alloys) were conducted in a relatively small number of studies.[5,6,69] However, examination of the process of formation of new phases at the diamond–melt interface and the structure and properties of these phases are important for explaining to the mechanism of these reactions, determination of the relationships governing the processes of wetting and adhesion and solving practical problems of metallising and brazing of diamonds and development of diamond tools.

The interaction of diamond with liquid nickel at 1500 °C and with a Ni–Si alloy at 1100 and 1500 °C was investigated in Ref. 70. A metal or alloy charge was melted on a diamond crystal and held for a specific period of time. Subsequently, sections were produced from the section of the droplet inclined in relation to the interaction plane. The sections were examined by metallographic techniques. The distribution of carbon and silicon in the section plane (size 2 μm, measurement error ±1.5%) was determined in a MAR-2 microanalyser. The results of X-

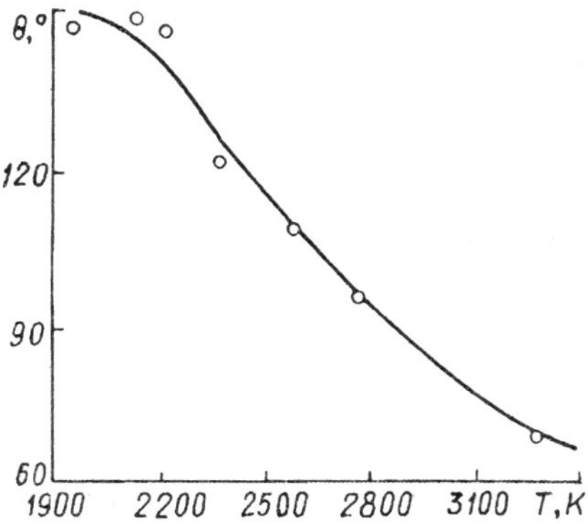

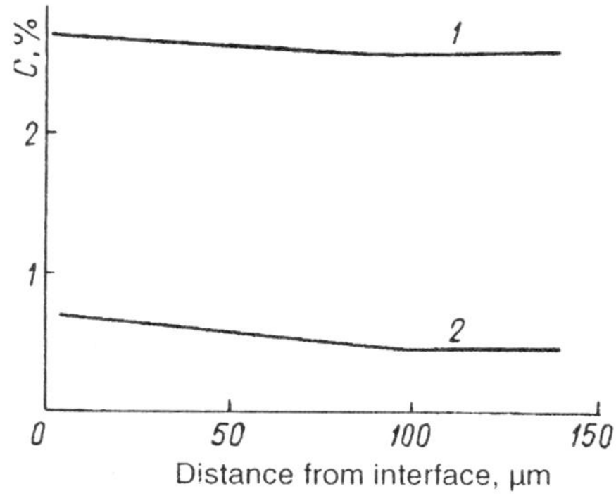

Fig.19 Temperature dependence of the contact wetting angle of copper on diamond at a pressure of 8 GPa.

Fig.20 Distribution of carbon in the coating after holding at contact with diamond (1500 °C, 15 min): 1) nickel; 2) Ni–Si alloy.

ray spectral microanalysis of nickel and the alloys based on Ni–Si, held for 15 minutes in contact with a diamond at 1500 °C, showed that diamond dissolves. Figue 20 shows the dependence of the carbon content in the solidified metal at different distances from the interface. The maximum carbon content in the boundary of nickel up to 20 μm thick is 2.7% which corresponds to its limiting solubility. However, the amount of carbon in the alloy in heating to 1500 °C is only 0.7% which is explained by a decrease of the solubility of carbon in nickel when silicon is added.

The authors of Ref. 71 and 72 examined the wetting of diamond by liquids metals and observed the formation and growth of an intermediate carbide layer.

The X-ray diffraction analysis of the composition of the new phase for the graphite–Cu–Cr alloy (5 %Cr) system, held in preliminary contact at 1200 °C in vacuum, shows that the phase consists of chromium carbides. The rate of growth of this phase was investigated in relation to the temperature–time conditions of the process and the activation (355.6 kJ/mol) and diffusion energy (209.2 kJ/mol) were estimated. The latter value corresponds approximately to the activation energy of carbon diffusion in the Cr carbide of 151.0 kJ/mol.

An intermediate carbide phase was observed in contact of other melts with graphite–germanium with tungsten and tantalum (carbides WC and TaC), and liquid uranium (UC_2 carbide).[39,73]

The thickness of the UC_2 layer, measured in relation to time, was relatively large and its growth, as assumed by the authors of Ref. 39, is associated purely with diffusion processes.

At the same time, the investigations carried out for titanium-containing melt with graphite[74,75] show that the layers formed during contact are very thin (less than 1 μm at 800–1100 °C at a titanium content of 1–4% in a copper or tin melt). The thickness of the intermediate layer for the Cr-, V- and Nb-containing melts under the same conditions is 5–10 μm.

The results show that the carbide layer always grows to the side of liquid metal.[76] Taking into account the structure of the resultant carbide layer and the process of carbon diffusion in it, it may be assumed that, regardless of the higher chemical affinity of titanium for carbon (ΔH = 188.3 kJ/mol of carbon) in comparison with, for example, Cr (ΔH = 48.5–68.6 kJ/mol of carbon), the rate of formation of the intermediate layer titanium carbide is lower.[77] This is associated with lower diffusibility of carbon in the layer of the thermodynamically stronger titanium carbide.

Later, the authors of Ref. 78 examined the chemical composition and structure of intermediate phases formed at the diamond-adhesion-active alloy interface by X-ray local microanalysis. Copper–titanium and copper–chromium alloys with an active metal concentration of up to 10% were selected. Cross sections of the contact zone were prepared after holding the metal in contact with diamond at 1150 °C for 45 minutes in vacuum and subsequent cooling. The composition of the transition zone between diamond and alloys was investigated using a Microscan-5 X-ray microanalyser. The measurement procedure made it possible to determine the content of titanium and chromium in the investigated specimens with a relative error not exceeding ±1.5%. The results are presented in Fig. 21.

A carbide interlayer with a zonal structure forms at the diamond –Cu–Cr alloy interface. Scanning the probe along the diamond-alloy line with recording of the intensity of CrK_α- and CuK_α-radiation in an automatic recording device, and also analysis of the carbide layer at different points confirmed the presence of zones with carbides of different type. The first zone, adjacent to the diamond, is the thickest (5 μm) and consists of Cr_3C_2 (85–87 % Cr). The second zone, of the same thickness, consists of Cr_7C_3 (89–90 % Cr). The presence of two intermediate zones can be clearly seen on microsections and the curves of distribution of chromium concentration in the thickness of the new phase (Fig. 21a). The third zone which is in contact with the alloy and is 0.5–1 μm thick can be seen on the microsection and in scanning of the probe is less distinctive than the other two phases. The authors and Ref. 78 assumed that this zone consists of the $Cr_{23}C_6$ (91–92 % Cr) carbide.

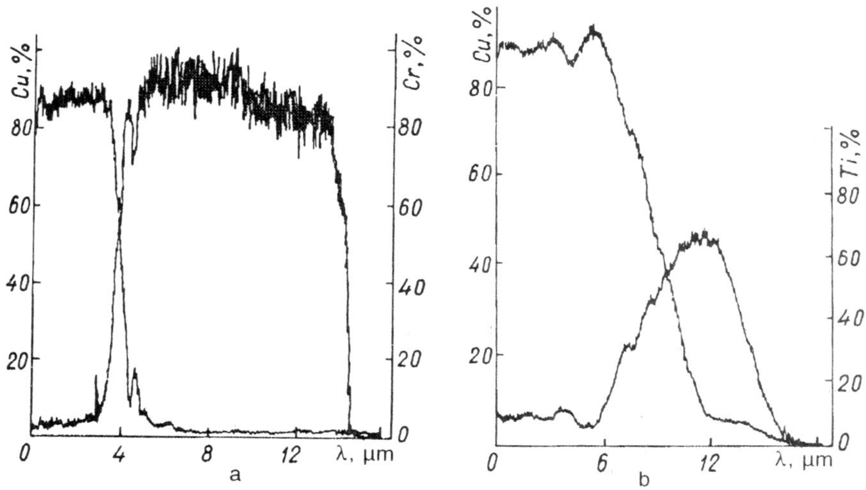

Fig.21 Dependence of the variation of the content of copper, chromium and titanium at copper–chromium alloy–diamond (a) and copper–titanium alloy–diamond (b) interfaces in transition from the alloy to diamond.

Thus, the intermediate phase, formed at the boundary between the diamond and the melt of the Cr-containing alloy, consists of three phases - Cr_3C_2, Cr_7C_3 and $Cr_{23}C_6$, which is in good agreement with the Cr–C equilibrium diagram and the data obtained in X-ray examination of Cr coatings on diamond[79] (due to small thickness, the $Cr_{23}C_6$ carbide layer was not detected by X-ray phase analysis).

Local analysis of the carbide layer at the interface of the diamond with the copper–titanium alloy showed (Fig. 21b) that the titanium content of this layer, 80–83 %, corresponds to the $TiC_{0.8-0.85}$ carbide.

The thickness of the titanium carbide is considerably smaller than the thickness of the chromium carbide layers.

Interesting results were published in Ref. 80. These results were obtained in examining boundary layers in the diamond–transition metal (titanium, chromium) system. Titanium or chromium specimens were brought into contact with diamond and heated in a vacuum of $1.3\cdot10^{-3}$ Pa in the temperature range 700–1000°C with holding for 30 minutes and two hours. The same compositions were subjected to the effect of high pressures ($70\cdot10^8$ Pa) at temperatures to 1200-1400°C. Examination of the interface in the diamond–titanium pair, compressed at a temperature of 1400°C and a pressure of $55\cdot10^8$ Pa (time 15 min), showed a transition zone consisting of the carbide layer adjacent to the diamond (4 μm), and a layer of the solid solution of carbon in titanium (10 μm).

The increase of temperature and pressure (1700°C, $65 \cdot 10^8$ Pa) results in a further increase of the thickness of the transition layer. The carbide layer is not saturated with carbon and its composition corresponds to $TiC_{0.8}$ at the interface with diamond. A transition layer 25 μm thick forms in the diamond–chromium couple compressed at 1500°C and a pressure of $60 \cdot 10^8$ Pa for 12 minutes. The results of X-ray spectrum microanalysis confirm that the first zone, adjacent to the diamond (4 μm), consists of Cr_3C_2 carbide, the second zone 20 μm thick corresponds to the Cr_7C_3 carbide and, finally, the third zone is the solid solution of carbon in chromium with the inclusions of the $Cr_{23}C_6$ carbide. This is in agreement with the previously mentioned data obtained in Ref.78, 79 without applying any pressure.

In Ref.81, the authors carried out experiments to examine the interaction of AC 15 diamond with copper melts containing 2–40% Ti, Cr, Mn, Fe, Co, and Ni in vacuum ($1–3 \cdot 10^{-3}$ Pa) and in purified hydrogen at 1150–1350°C. The composition of the newly formed phases was also examined using a Microscan-5 microanalyser, and a procedure described in Ref.78; the probe diameter was 0.3–0.8 μm. This detailed examination enabled the authors to determine reliably the phases formed at the interface with the melts.

The formation of a dense layer of the TiC carbide at the interface with diamond has been confirmed for copper–titanium melts (5–10% Ti). Interaction of synthetic diamond with the Cu–Cr melt (2–5% Cr) is characterised by the formation of a layer of Cr carbides at the interface. The data for the kinetics of growth of this layer are represented in Fig. 22 and are close to the values obtained in Ref. 5.

According to the results of local analysis, the carbide layer consists of three zones – Cr_3C_2, Cr_7C_3 and $Cr_{23}C_6$ carbides, which is in agreement with results published in Ref. 79.

The authors of Ref. 80 also confirmed the same distribution of the carbides in the contact zone in the solid-phase interaction of diamond

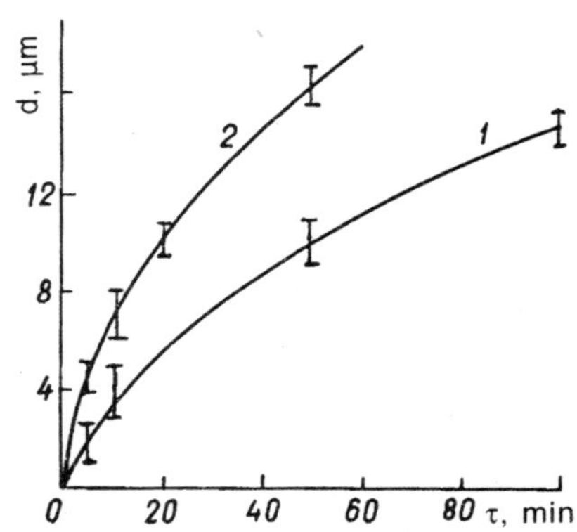

Fig.22 Kinetics of the growth of the carbide layer in interaction of synthetic diamond with Cu–Cr (5%) melt at 1150 (1) and 1200°C (2).

with Cr. They reported several special features of the formation of Cr carbides on synthetic diamond crystals. For example, the carbide layer formed of the (111) faces is usually slightly thicker (15–20%) than on the (100) faces. The Cr carbides formed not only on the crystal surface but also in the cracks splitting the crystals along the (111) planes. In the cracks, whose width reaches the 20–25 μm, in addition to chromium there is also copper, but in thin cracks copper was not detected by X-ray local analysis which indicates the selective adsorption of Cr on the diamond surface (Fig. 23).

The interaction of diamond and manganese was investigated using a Cu–Mn (40%) melt. The results show that at 1150 °C, 10 min, a carbide layer 3.5 μm thick forms at the diamond–melt interface, whereas at 1200 °C the thickness of the layer in the same time reaches 8–9 μm. The boundaries of the carbide layer with both diamond and the melt are unsharp, uneven, and there are projections of individual crystals of manganese carbide growing toward the melt.[81]

The distribution of copper and manganese in the contact zone of the alloy with diamond is shown in Fig. 24. The manganese content

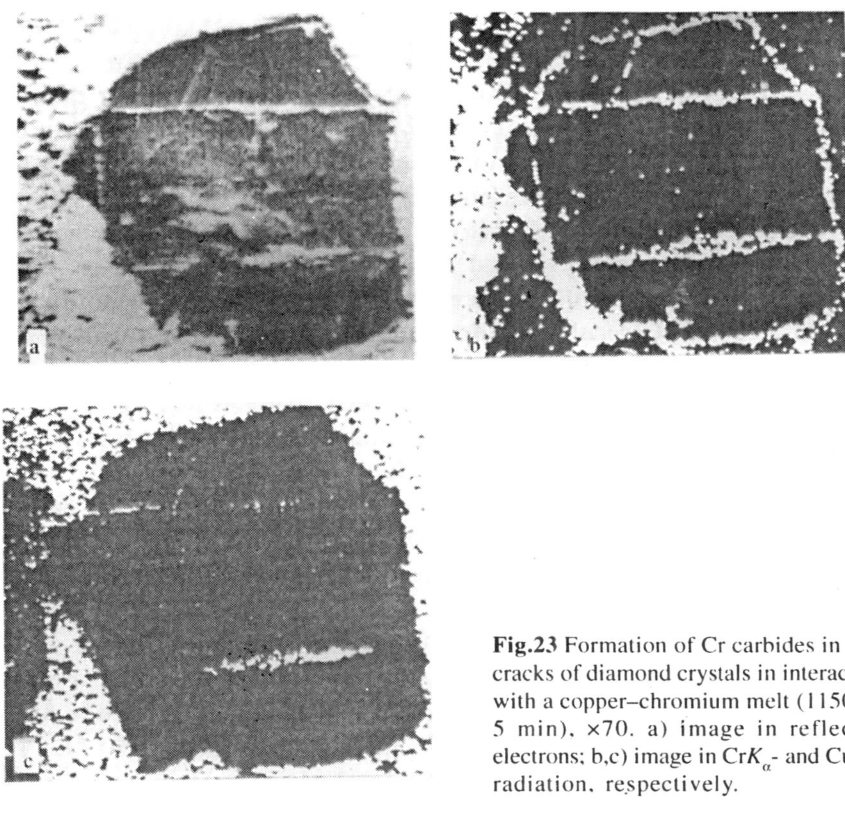

Fig.23 Formation of Cr carbides in fine cracks of diamond crystals in interaction with a copper–chromium melt (1150°C, 5 min), ×70. a) image in reflected electrons; b,c) image in CrK_α- and CuK_α-radiation, respectively.

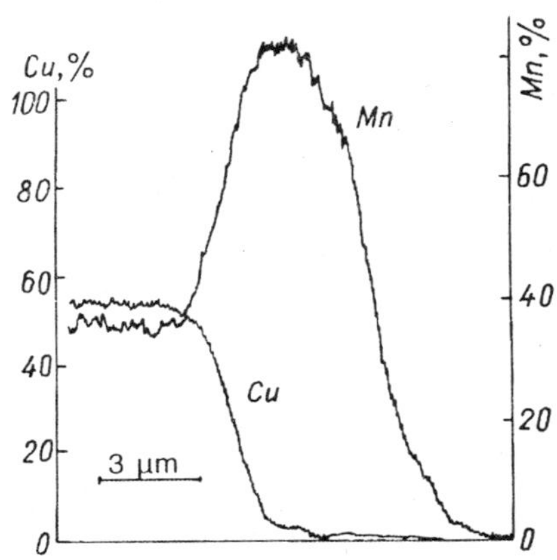

Fig.24 Distribution of manganese and copper at the contact of diamond with the Cu–Mn (40%) melt (1150°C, 10 min).

of the newly formed phase reaches 82–83%, the copper content 2–3%. It can therefore be assumed that the layer contains carbon-rich carbides. X-ray diffraction patterns of the diamond crystals, coated with a carbide film, contained diffraction lines corresponding to diamond and manganese carbides Mn_5C_2 and Mn_7C_3.

At the interface of diamond with the copper melts, containing iron, cobalt, nickel, new phases do not form and the diamond crystals rapidly dissolve with the precipitation of finelly dispersed graphite. The rate of dissolution of synthetic diamond in these melts is higher than that of natural diamond. This is caused by the presence of impurities in their lattice and also of inclusions of metals used in synthesis.

Interesting results were presented in Ref. 82 in examining the contact interaction of graphite in Ni–Mn–Ti and Ni–Mn–Sn–Ti melts using X-ray spectrum microanalysis and X-ray diffraction analysis.

Investigations carried out on a tin-free alloy show that holding for 15 min at 1050 °C results in complete saturation of the entire volume of the alloy with carbon, and carbide inclusions were detected at large distances from the alloy–graphite interface. The carbide layer was not detected at the interface. Evidently, this was caused by higher solubility of carbon in the Ni–Mn–Ti melt.

Thus, the chemically active elements (with high affinity for carbon), for example, titanium, are not adsorbed at the interface, as observed for many other systems with titanium described in the literature (Cu-Ti, Sn–Ti, etc).

The addition of tin which greatly reduces the solubility of carbon in the alloy changes the course of the reaction at the alloy–graphite interface. In particular, a continuous layer of manganese carbide Mn_7C_3 (Fig. 25 a,b) forms at the interface in the range 900–1050 °C. This layer is followed by an alloy with the microstructure typical of the deep

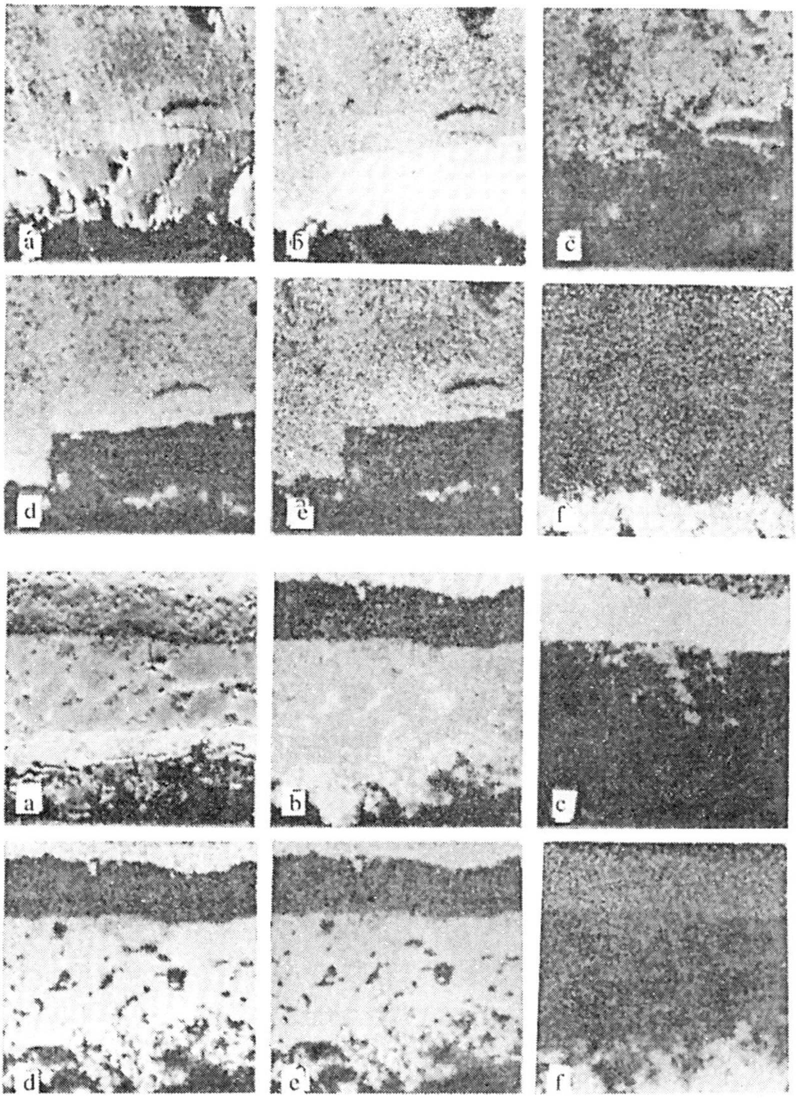

Fig.25 Contact zone of graphite–alloy 70 (Ni–Mn)–20 Sn–10 Ti system after heating to 1050°C with holding for 15 min (upper photograph) and heating to 1100°C and holding for 30 min (lower photograph), ×500. a) image in reflected electrons; b) in characteristics rays MnK_α; c) TiK_α; d) NiK_α; e: SnK_α; f) CK_α. The upper part of each photo is the metal alloy, the lower part is grahite.

zones away from graphite: grains of the Ni_3Mn_3Sn ternary compound and the eutectic. The titanium carbide is found at the interface in the form of island inclusions, Ni, Sn are absent (Fig. 25 d,e).

The increase of temperature complicates the nature of the chemical reaction. For example, at 1100 °C (holding time 30 min) a diffusion layer consisting of three sublayers appears. In the immediate vicinity of graphite there is a Mn_7C_3 sublayer followed by the Ni_3Mn_3Sn intermetallic compound and, finally, the TiC layer (Fig. 25 a,b,c). This complicated redistribution of the elements and also the order of distribution of the phases in the diffusion layer are caused by the features of the reactions taking place in the contact zone – the transfer into the alloy of carbon thermodynamic activity of metallic components in the melt, diffusion of components in the solid and liquid phases, the kinetics of interfacial interactions. Here reaction Mn + $\frac{3}{7}$C → $\frac{1}{7}Mn_7C_3$ is the main process of interaction of carbon with the alloy (Ni–Mn–Sn–Ti).

The authors of Ref. 83 examined the contact interaction of the surface of a carbon material with Al-based melts containing 6% Sn and a variable amount of titanium: from 0.5 to 10%. The results showed, Fig. 26, the very strong effect of Ti on wetting.

These results were explained on the basis of the results of X-ray spectrum microanalysis of the transition layer. The results show that at a temperature of 1220 K up to 60% Ti and 17% Sn concentrate in the zone of contact of the liquid phase with the solid substrate. The nature of distribution of the intensities of the characteristic radiation of titanium and tin at the examined points indicates that part of titanium in the transition layer is in the form of a compound of titanium with tin whose stoichiometric composition approximately corresponds to the Ti_3Sn compound.

X-ray diffraction analysis of the examined specimen confirmed the presence of Ti_3Sn compound in the transition layer. This analysis also showed the presence of a titanium carbide layer on the surface in the immediate vicinity of graphite.

The formation of the Ti_3Sn compound in the transition layer depends on the titanium content of this layer. Taking into account the equilib-

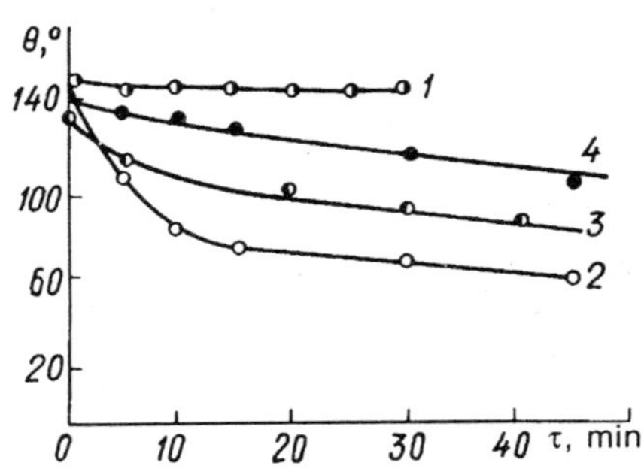

Fig.26 Wetting of the surface of the carbon material by melts based on aluminium (Al–6% Sn) – Ti at a titanium content of: 1) 0.5; 2) 1; 3) 3; 4) 5%.

rium diagram of the Ti–Sn syste..., the titanium content required for the formation of this compound should be higher than 50%. In this case, up to 60% Ti is concentrated in the transition layer at 1220 K, i.e. the amount sufficient for the formation of the Ti_3Sn compound.

Thus, the processes of interaction of diamond (graphite) with carbide-forming metals in the solid state and with melts containing additions of these metals are as a rule accompanied by carbide formation at the interface; the intensity and rate of growth of the carbide in the initial stage are determined by the kinetics of chemical reactions at the interface and then by the diffusion of components (carbon and metal) through the layer of the new phase. The second stage is usually controlling. However, in cases in which the temperature is high and the contact time is short, the 'reaction' stage of the process can play a significant and, possibly, controlling role.[6]

1.3 INTERACTION OF DIAMOND WITH METALS IN DEPOSITION OF COATINGS

The methods of depositing coatings on the diamond were examined in detail by the authors of this book in Ref. 1. In this monograph, the currently available methods of metallising diamonds are described briefly.

The process of metallising diamond and other superhard materials is based on the phenomenon of contact interaction of metals and alloys with their surface.

The results of investigations in the area of interaction of the melts with the diamond surface (graphite) and the work in the development of the theory of wetting, diffusion and contact phenomena in the diamond-metal system make it possible to examine the problems of metallising diamond and select coating composition and conditions.[1,6,85] This region of investigations is being rapidly developed and new methods of metallising and coating compositions have been proposed. At present, the following methods are used most widely:

1) Chemical and electrochemical methods of depositing coatings from solutions and melts of salts;

2) The methods of gas-transport reactions (deposition of metals from gaseous silanes, halides, carbonyls);

3) Deposition of coatings from a liquid metal melt;

4) Beam methods (plasma, cathode, electron-beam sputtering);

5) Mechanical cladding of grains;

6) Contact-reaction deposition of coatings on diamonds from mixtures with a powder metallising agent.

As shown in the Ref. 86, two main problems must be solved when depositing coatings: the formation of a uniform metallic film on the surface of the abrasive grain; the development of adhesion at the film-substrate interface. To satisfy the second condition, it is necessary to select a metal with a high chemical affinity for carbon, and also provide conditions under which chemical bonds can form. The methods of depositing coatings on diamonds will be briefly discussed.

Chemical and electrochemical methods

The electrochemical method is used in galvanic systems with the passage of electric current through an electrolyte solution based on the salt of the metal to be deposited on diamond.[87,88] The process includes heating to 70-80 °C with agitation of the electrolyte. Copper, nickel and cobalt coatings produced by these methods are used in practice, although coatings of other metals (Pt, Mo, Ta, Nb, Ti)[89-91] are also available together with coatings of alloys (Fe–Ni, Ti–Co).[92] The electrochemical method of metallising diamond is used widely by foreign companies (Norton, ASEA, De Beers).[91,93]

The thickness of deposited coatings is regulated by the duration of the process, solution concentration, electrolyte temperature and the current density in the electrolyte. To accelerate the process of electrochemical deposition of coatings, it is recommended to use ultrasound with a frequency of 800–3000 kHz.[94] The authors noted better adhesion of the metallic coating to the abrasive grain and acceleration of the coating process.

The diamond grain, metallised by the galvanic method, is enclosed in a metal shell which is chemically not bonded with the diamond surface because the coatings are deposited at low temperatures when there is no formation of chemical bonds. The results of examination of the strength of contact of these coatings with the diamond show that the strength does not exceed 1 MPa. This is a serious shortcoming of this process.

As regards chemical deposition of metals from the solution of the salts on diamond, it should be noted that the deposition of metals is also carried out at moderate temperatures with constant agitation and compulsory treatment of the surface (cleaning, degreasing, sensitising the diamond surface by treatment in a solution of $SnCl_2$ or $PdCl_2$) to intensify the process. The mass of the coating is determined by weighing after operation. The process is difficult to control when it is required to deposit a specific mass of the coating. In most cases, this method is used to deposit Cu, Co, Ni.[95, 96] The authors of Ref. 97, 98 proposed the deposition of tungsten, molybdenum, titanium and other coatings.

The gas transport reaction method

This method is used to coat diamond with a number of carbide-forming metals, such as silicon, titanium, tantalum, tungsten and other metals using silanes of silicon, halides (iodides) of titanium and tantalum in interaction with hydrogen. For example, the silicon coating on diamond produced by the methods described in Ref. 99 forms in interaction of methylchlorosilane with hydrogen at a temperature of 1325 °C, the tantalum and titanium coatings form i dissociation of hydrides in a zone with a temperature of 1000°C which contains [100] diamonds, and tungsten coatings formed at temperatures of 1000–1050 °C resulting the formation of carbide layers at the interface and, consequently, sufficiently high adhesion.[101] The thickness of coatings formed under these conditions is 2–5 μm.

A method similar to that described above is the method of metallising diamonds based on thermal dissociation of carbonyls of tungsten, molybdenum, iron and rhenium. The coating deposition temperature depends on the metal used: for iron it is 130 °C,[102] for tungsten and molybdenum 300–500°C,[103] although high temperatures are also possible and, consequently, an intermediate carbide layer can form at the interface. The rate of deposition of these coatings is usually 1–1.5 μm/h.[104] The bonding strength of these coatings with diamond is usually low (1–3 MPa).

However, it should be noted that the gas-transport reaction method, including the carbonyl method, requires special equipment, strict inspection of the gas phase, the components of the reaction gas mixtures are often toxic and explosive. This greatly restricts the application of this method under industrial conditions.

Deposition of coatings from a liquid metal melt

This process is based on liquid-phase sintering of the metal of the coating with diamond. For this purpose, a charge of metallic powders (component of the coating) is mixed with diamond powder and placed in a vacuum furnace (under the conditions of high vacuum because the metallic composition contains adhesion-active additions), and is then held at the melting point of the metal of the mixture for the required period of time. The resultant cakes are then crushed and screened. The mass of the coating is controlled by the loaded mass of the metallic charge.[105] The produced metallised diamonds have the form of diamond grains or aggregates of diamond grains surrounded by the metal shell.[106] The special feature of the vacuum method of metallising is the strong chemical bonding of the coating with the diamond surface as a result of the formation of the boundary carbide layer. In addition, under the effect of capillarity forces the metallic melt fills the smallest pores, cracks and defects on the diamond surface and, after hardening, has a unique case-hardening effect which increases the strength of the diamond grain 1.4–2 times.[107]

The diamonds metallised by this procedure are placed in a meta 'jacket' which protects them against pulling out and in the case of cracking against the release of fragments from the tool.

If set and operated properly, vacuum equipment can provide high productivity and satisfies the requirements of industrial manufacturing.

A typical example of the vacuum method is the technological process of metallising and aggregation of diamonds by Cu–Sn–Ti alloys.[108]

The liquid phase method of metallising diamonds was then used as a basis for developing the process of hardening micropowders using the Ni–Si–Sn composition.[109] The vacuum metallising method was proposed in the Ref.110, 111. In this method, a layer of powder transition metals, for example, titanium, zirconium or their alloys, is deposited on the surface of the diamond particles. The powders are then heated in a vacuum of 10^{-2} Pa to 350–1000 °C (optimum temperature 900 °C) for 10 min and cooled; this results in the formation of single grains with a thin coating (of uniform thickness) of transition metals with a strong bond at the metal–diamond interface as a result of the formation of the carbide phase.

If necessary, the particles can be coated by several layers of metals with lower oxidation capacity, for example, with successive layers of copper and nickel. The particle can contain projections or be coated by sponge iron, for instance.

The deposition of coatings by liquid-phase sintering in inert and reducing media has been examined in Ref. 112. In the process, coatings are deposited on diamond in the liquid phase in the inert medium by heating to the melting point of the metal. The composition of the coating can include highly volatile components of the metals (for example, antimony). Ar or N can be used as the inert medium. Some authors believe that the inert medium is most suitable for sintering diamonds with the coating. However, this has not been confirmed in practice. The productivity of this method is quite high, although it is necessary to take into account the high cost of the inert gas and the complicated operation of cleaning systems for inert gases.

Diamonds are also metallised in reducing media. For example, the authors of Ref.112 proposed to metallise diamond in hydrogen at 540–650 °C.

In Ref. 85, a coating was produced on diamonds by introducing diamond powder into a reactor through a nozzle by the flow of fuel gas and by the formation of a fluidised layer into which the metallic melt was added in the form of aerosol.

On the whole, the method of depositing coatings from the liquid phase, especially with adhesion-active components taking part, has a number of advantages in comparison with other methods and results in high bonding strength of the coatings with diamonds and formation of coatings of the required thickness. Simple technology and apparatus for metallising enable this method to be used widely in industry.

Beam coating methods

Coatings are also deposited from a flux of atoms or ions of the metal produced using different sources. For example, a jet of ionised gas (plasma) is used in the plasma method.[113, 114]

A special feature of the plasma jet is the very high velocity of the particles forming this jet. This velocity depends on the voltage between the electrodes. In the process of formation of the plasma coating it is possible to deposit any refractory materials with high speed and uniformity, including tungsten, molybdenum, carbides, oxides, borides, etc. The process takes place in Ar or in vacuum (approximately 10^{-2} Pa).

According to Ref. 114, when the plasma jet and microparticles hit the surface to be deposited, their kinetic energy generates a pressure of the order of 1–100 MPa in the collision zone. These high pressures, generated in the contact zone, should, according to the authors, result in the formation of a chemical bond at the metal–diamond interface, especially when using a carbide-forming metal. The thickness of the coating is regulated by changing the voltage between the electrodes and the process time. However, it should be noted that the ionisation of atoms and the collision of the atom with the hard surface can evidently activate kinetically the formation of chemical bonds but only if the atom has high chemical affinity for the given surface, for example, diamond. If there is no such affinity, chemical bonds not formed and the bonding strength at the metal–diamond interface is low. This was also observed in Ref. 114 where the authors obtained the sufficiently high strength of the cubic boron nitride and diamond polycrystals with a Cu–Ti coating (70 and 60 MPa, respectively) and low strength with a copper coating – to 10 MPa.

According to the metallising method proposed in Ref. 115, the metal to be deposited evaporates in a vacuum of 10^{-3} Pa as result of electron beam heating. The adhesion of the coating is determined here by the temperature to which the substrate (diamond) is heated. For example, when spraying molybdenum films by the electron-beam method on the diamond heated to 800 °C, the bonding strength was, according to the data in Ref. 116, approximately 2 MPa, and at higher temperature the bonding strength was considerably higher.

In Ref. 117–119, the authors proposed a method of cathode sputtering of metal for depositing Ag-, Ni-, Cu-, Pt-, W, Co and Ti-coatings. The results show[119] that titanium is efficiently sputtered in the glow discharge and forms thick coatings on diamond. The temperature of the zone in which the diamond is situated has also a significant effect.

The authors of Ref. 120 examined the dependence of the process of metallising and adhesion of films of titanium, chromium and nickel with single crystals of synthetic diamonds on the parameters of glow discharge in spraying (sputtering voltage, time and current density). The results show that at a fixed sputtering time, with other conditions being equal, thicker coatings were produced in the case of nickel, whereas the titanium coatings have the smallest thickness, Fig. 27. Evidently, this can be explained by the difference in the sputtering coefficients of these metals.[121] The strength of the bonding of the films of the examined metals titanium and chromium with diamond, determined using the procedure described in Ref. 122, after heat treatment at 500–600 °C for 30 minutes was 40–50 MPa, and the same strength

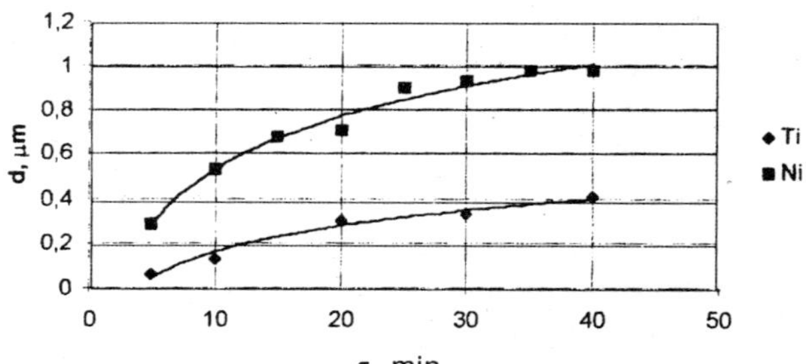

a

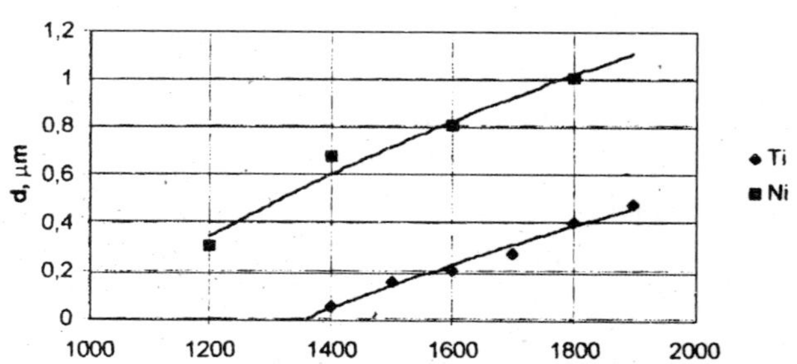

b

Fig.27 Dependence of the thickness of films of metals Ti and Ni on sputtering time (U_p = 1600 V, I = 4–5 mA/cm^2) (a) and sputtering voltage U_p (τ = 30 min, I = 4–5 mA/cm^2) (b).

values were recorded after preliminary plasma treatment of the crystals of synthetic diamonds prior to deposition.

Because of the fact that equipment used is complicated, these methods are still used only under laboratory conditions.

Mechanical cladding

In this method, the coating can be deposited by rolling diamond grains wetted by a volatile solvent, in metallic powders, with subsequent heat treatment to increase the adhesion of the coating to the surface of the abrasive grains.

This method was used, in accordance with a patent described in Ref. 123, to coat diamond with powders of iron, cobalt and nickel and also WC and their mixtures. A similar method was used in Ref. 124 in depositing carbides of tungsten, tantalum, titanium and chromium on diamond. For example, this method was used to deposit titanium on diamond grains,[125] which were then annealed in a vacuum of 10^{-3} Pa at temperatures of 850–900 °C resulting in strong bonding with the surface with the formation of TiC at the interface.

To increase the thickness of the metallic coating, in Ref. 126 nickel was initially deposited by electroplating on diamond powder and then rolled in carbonyl nickel powder mixed with, for example, 12% solution of rubber in petrol. After granulating the charge by rubbing it through a sieve with a specific mesh size followed by sintering, granules of diamond particles with the nickel porous coating were produced. The porosity depends on the sintering conditions. However, this method is not used often because it requires intermediate operations and heat treatment to increase the adhesion strength of the coating with the substrate. Consequently, the method is not yet used in industry.

The contact-reaction method

A special group includes the methods of coating diamond with coatings in which the metallising process takes place in mixtures with powder-like metallising agents in the solid phase. This method was developed for the first time for metallising diamonds with metals characterised by high vapour tension, especially chromium.[127] A variety of this method is the procedure proposed in Ref. 128 whose special feature is that to accelerate the metallising process and produce thick coatings, an activating agent is added to the powder of the metallising agent — the fluoride of an alkali or alkali-earth metal, in an amount of up to 1.5%.

Recently, a method has been developed of coating diamond powders with refractory carbide-forming metals (tungsten, molybdenum, titanium, chromium)[129] at temperatures of 800–1000 °C. In this case, the metallising agent is represented by surface-oxidised powders of refractory metals.

This method is highly promising for developing industrial technologies. It is characterised by high productivity, is simple as regards technologies and apparatus, is characterised by high adhesion of the coatings to the diamond, and can be used also in metallising synthetic diamonds with low heat resistance, i.e. ballas, carbonado.

The authors of Ref. 130 proposed a method of depositing coatings on diamond including thermal diffusion saturation in a rotating container. For this purpose, a charge consisting of powders of diamond, the metal (the material of the coating) and ammoniaum chloride NH_4Cl, is loaded into a container placed in a silit-rod electric furnace in which the diamond particles are metallised. As a result of rotation, non-stationary gas flows are generated in the container, the rate of removal of reaction products is increased, the powders loosen up so that the diamond particles become accessible for depositing coatings. The experimental results show that at a frequency of rotation of the container of 40 min^{-1}, the degree of filling of the container of 25–30%, the volume fraction of diamonds in the charge of 20–25% and the NH_4Cl content of the charge of 2–3%, it is possible to deposit coatings of the maximum thickness on diamonds. In particular, to deposit titanium and chromium coatings on diamond particles ~300 μm in size, the authors recommend the following technological parameters: t = 700 and 800 °C, τ = 60 min, the ammonium chloride content 2.5%.[131] This is characterised by the maximum strength of metallised diamonds (Fig. 28a). The increase of temperature and process time (Fig. 28b) results, according to the authors, in dechlorination, i.e. displacement of gas-transport reactions to the initial conditions. The authors examined in detail the effect of this method of depositing coatings on the strength of metallised diamonds with various numbers of defects in the structure, carried out x-ray phase analysis of the coatings and published rec-

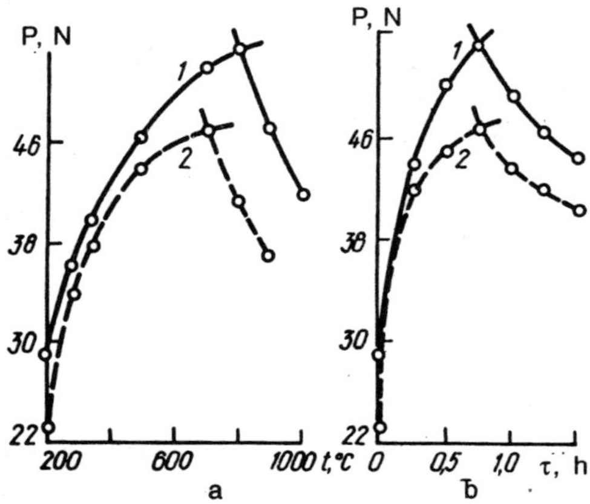

Fig.28 Dependence of the fracture load (strength of 315/250 μm diamond powders, metallised with chromium (1) and titanium (2) on temperature (a) and process time (b).

ommendations regarding the effect of titanium and chromium coatings on both the strength of diamonds and their interaction with the binder of the tool where such coated diamonds were used. The efficiency of tools was also investigated.[132]

The analysis of these metallising methods shows that they have certain advantages and disadvantages.

A specific metallising method should be selected in order to obtain coatings of different composition, density, thickness, and the bonding strength with the substrate. For example, the coatings deposited on diamond by electrochemical deposition are characterised by low contact strength. Metallising of diamonds by the beam method and by the method of gas transport reactions results in sufficiently high bonding strength but the complicated nature and high cost of equipment do not yet make it possible to use these methods in practice.

The following methods are most promising for practice: 1) deposition of coatings from the liquid adhesion-active metallic phase because this method has significant advantages — strong bonding of the coatings with the diamond surface as a result of the formation of the carbide interlayer, the possibility of regulating the thickness of the coating, deposition of the coating with the required properties; 2) the contact-reaction method of depositing coatings of refractory carbide-forming metals resulting in the formation of uniform coatings with high adhesion to diamond.

These two methods are characterised by high productivity and technological features so that they can be used widely in industry.

Special attention has been given to examining the formation of carbide layers in depositing on diamond coatings of transition metals - titanium, vanadium, niobium, tantalum, chromium and molybdenum.[133-140]

The powder of natural diamond with a grain size of 63/50 μm was metallised with titanium, vanadium, chromium and molybdenum in vacuum heating in the powder-metallising agent using the method described in Ref. 127, 129. The metallised diamond powders were investigated by recording and decoding x-ray diffraction patterns in a Debye-type chamber (diameter 0.15 m) in chromium and copper radiation. References were represented by x-ray diffraction patterns produced for pure metals and also carbides of different composition.

Indexing of the x-ray diffraction patterns was carried out using the information presented in Ref. 141 and 142.

Niobium and tantalum coatings were deposited on natural diamond powders with a grain size of 400/315 μm in vacuum annealing in a metallising agent consisting of a mixture of surface-oxidised powders of niobium and tantalum and also NaF activator (1–4%) by the method proposed in Ref. 143. In the metallising diamonds by these metals without the activating agents the coating did not form even at relatively high temperatures (1100–1200 °C). Evidently, this is explained by the low tension of the vapour of niobium and tantalum oxides.[144,145]

The effect of the activator in the metallising process is based on the formation, as a result of chemical reactions, of highly volatile compounds – oxyfluorides of niobium and tantalum which, evaporating and depositing on the surface of diamonds, are then reduced by carbon to the metallic phase.

X-ray examination of niobium and tantalum coatings on diamond powders was carried out in DRON UM-1 equipment in monochromatic radiation CuK_α by the method of x-ray diffractometry. The experimental data were processed in a Mera computer using standard programmes.

The results of x-ray diffraction analysis of the titanium coatings show that at 800°C (τ = 1 hour) a coating layer (approximately 0.4 μm thick) forms on the diamond and yields weak reflections from Ti and very weak reflections from TiC. The thickness of the coating increased with increasing temperature. The x-ray diffraction patterns show an increase of the intensity of TiC lines and a decrease of the intensity of titanium lines. At temperatures above 950°C (τ = 1 hour) there are no titanium lines on the x-ray diffraction patterns and the coating consists almost exclusively of the TiC carbide.

The effect of metallising time on the kinetics of formation of the titanium coating has been examined in detail at 900°C. The results show that after 15 minutes holding a thin coating (approximately 0.1 μm thick) forms on the diamond. On the x-ray diffraction patterns, this coating produces weak reflections from titanium and TiC. When the metallising time is increased to 1 hour the thickness of the coating increases. This is accompanied by an increase of the TiC content of the coating and a decrease of the titanium content. With time there is a rapid decrease of the rate of growth of the thickness of the coating and the titanium lines disappear from the diffraction patterns. A similar trend was also detected on x-ray diffraction patterns taken from diamond powders after their high-temperature metallising (950, 1000 °C) with different holding time.

In the initial metallising stages (τ = 0.25-0.5 h, at temperatures of 900, 950, and 1000 °C) the titanium carbide is characterised by a low value of the constant of the cubic lattice. This may indicate that its composition differs from stoichiometric and its lattice contains carbom vacancies.[146,147] With increase of metallising time to one hour the carbon vacancies are filled and the lattice constant of the TiC increases. A further increase of the metallising time (τ = 1 h, t = 900 °C) does not cause any large changes in this parameter. In the case of one hour metallising an increase of temperature from 900 to 1200°C (within the limits of the measurement error) doesn't cause any large change of the lattice constant of TiC.

The chromium carbides Cr_7C_3, Cr_3C_2 and Cr were found in chromium coatings on diamond.[136] X-ray examination of metallised powders, etched in sulphuric acid (T = 50-60 °C, τ = 10 min) showed that these phase constituents are not distributed uniformly in the coating but concentrate in the coating in layers: in contact with diamond Cr_3C_2 and then Cr_7C_3 and chromium on the surface. The phase composition of the chromium coatings is strongly influenced by the temperature of the metallising process of diamond. For example, at 800 °C a thin layer of the coating characterised by weak reflections from Cr, Cr_7C_3 and Cr_3C_2, forms on its surface. When the temperature is increased to 1000 °C, the thickness of the coating increases and, at the same time, the content of the carbide phase increases because of activation of carbon diffusion.

The temperature range 1000-1050 °C is characterised by the rapid growth of the thickness of the coating and carbide layers (the intensity of the lines, especially of Cr_7C_3 on the diffraction patterns increases). At the same time, the amount of chromium in the coating decreases and this may indicate that the rate of carbide formation is higher than the rate of deposition of chromium. When the metallising temperature is increased to 1200 °C, the intensity of carbon diffusion becomes so high that a large amount of carbides (mainly Cr_7C_3) already forms in the metallising agent. These carbides inhibit the supply of chromium to the surface of the diamond grain. This reduces the rate of increase of the thickness of the coating and leads to saturation of the coating with the carbides.

It is characteristic that under these metallising conditions graphite is found at the interface with diamond. As regards the time dependence of the metallising process, the following conclusions can be made. At 1000 °C after holding for 15 min a thin film consisting of chromium and a small amount of Cr_3C_2 forms on the diamond. An increase of the metallising time to one hour results in an increase of the thickness of the coating in which the content of the carbides, both Cr_3C_2 and Cr_7C_3, increases.

A further increase of the metallising time greatly reduces the rate of growth of the thickness of the coating (as a result of the reduction of the rate of carbon diffusion in the growing carbide layers and, possibly, as a result of exhaustion of the metallising agent) and decreases its chromium content. As regards the carbide phases, the preferential growth of the higher carbide Cr_3C_2 in comparison with Cr_7C_3 becomes evident.

At 1200 °C the rate of diffusion processes is higher and, consequently, the resultant coating is so thick that it completely absorbs the reflections from diamond. After metallising for two hours, examination by the x-ray diffraction method did not show any chromium in the coating, and after 5 hr the coating consisted almost exclusively of the higher carbide Cr_3C_2. Graphite lines appeared on the x-ray diffraction patterns.

Similar investigations of the contact interaction of diamond with chromium[148] have confirmed the presence of Cr_3C_2 and Cr_7C_3 carbides on the interface. The $Cr_{23}C_6$ carbide was also found.

The results of analysis of the experimental data revealed a number of processes, often interlinked, comprising a complicated mechanism of metallising diamond with chromium, with different temperature and time dependences of the kinetic parameters. These processes include the evaporation of chromium from the metallising agent and deposition of chromium on the surface of diamond or the coating, diffusion of carbon (and, possibly, of chromium) in the carbides and chromium, and the formation of carbides with different carbon content ($Cr_{23}C_6$, Cr_7C_3, Cr_3C_2), in both the coating and the metallising agent. In addition to these processes, graphitisation of diamond also takes place at high temperatures.

Thus, the phase composition of the chromium coatings on diamond changes in relation to temperature and metallising time and, as shown previously, the structure of the coatings may consist of layers changing from Cr_3C_2 in the vicinity of diamond to chromium on the surface of the coating. With increasing temperature and holding time in metallising the thickness of the chromium layer decreases and the carbide layers grow up to the formation of the coating consisting almost completely of the carbides.

In the case of metallising diamond with molybdenum, a thin coating forms at relatively low temperatures (800°C). This coating consists mostly of the concentration-nonuniform carbides Mo_2C and a solid solution, evidently of carbon in molybdenum, Mo(C). The coating deposited at 900-950°C consists of homogeneous carbide Mo_2C and pure molybdenum.[137]

A further increase of metallising temperature supports the increase of the thickness of the carbide interlayer distributed, as shown by layer analysis, directly on the diamond surface. The thickness of the metallic phase remains almost unchanged.

Increasing metallising time at 950°C initially results ($\tau \leq 0.5$ h) in the formation of a thin film consisting of molybdenum and Mo_2C carbide. In subsequent stages, the thickness of the coating increases as a result of increasing amount of the carbide phase.

In Ref. 149 in investigations of the contact zone formed as a result of the interaction of diamond with silicon and tungsten at 900–1200 °C, examination showed carbides SiC, WC and W_2C when the temperature exceeded 1100 °C. The increase of temperature above this level resulted in graphitisation of diamond. The authors assumed that the process is accompanied by the rupture of C–C bonds in diamond and takes place through an intermediate stage of the amorphous layer which is a disordered configuration of the carbon atoms which rapidly condensed with the formation of the stable structure of graphite.

The results of examination of the zone of contact of diamond with vanadium coatings[138] showed that the coating deposited at 900 °C consists of V_2C and VC, at 1000 °C of V, V_2C and VC, and at 1100 °C of VC. It is probable that in the initial stages of metallising the interaction of diamond with the oxides results in the formation of the lower carbide V_2C on the surface of the diamond. With time, as a result of diffusion V_2C is saturated with carbon and a layer of VC forms at the interface with diamond. At 900 °C, according to the lattice spacing (a = 0.4137 nm), the composition of VC is close to the lower boundary of homogeneity.[150, 151] When the temperature is increased to 1000 °C, carbon diffusion is activated and, consequently, the composition of vanadium monocarbide approaches the upper boundary of homogeneity (a = 0.4166 nm). With a further increase of temperature the diffusion of carbon from diamond to the external surface of the coating is equalised by the deposition of vanadium. The coating is saturated with carbon, and at 1100 °C after metallising for one hour the coating consists completely of the vanadium monocarbide.

Niobium coatings start to form on the surface of diamond at 800 °C. A thin coating (h = 0.05 μm), deposited at this temperature, consists of approximately identical amounts of the metallic and oxide phases. With increasing metallising temperature and activator concentration the thickness of the coating increases and its phase composition changes: the thickness of the carbide increases and the thickness of the oxide and metallic phases decreases. At temperatures of 1100 °C and higher (3% NaF), when the thickness of the coating exceeded 0.8 μm, the coating consisted almost completely of the carbide phase (Table 5).[139] The lattice constant is 0.4462 nm, which is in agreement with the literature data (0.4449 nm).

Tantalum coatings, in contrast to titanium, vanadium, niobium, chromium and molybdenum coatings, start to form at higher temperatures (900-950°C). The variation of the activator concentration in the range 2-4% has no influence on the thickness of coatings deposited at constant temperature. The thickness of the coating on diamond increases with increasing temperature most rapidly in the temperature range 1000–

Table 5 Phase composition of niobium coatings on diamond, %

Phase	Metallising temperature, °C (3% NaF)			
	800	950	1100	1300
Nb_2O_5	40 – 45	0	0	0
Nb	55 – 60	20 – 25	0	0
NbC	0	75 – 80	100	100

1200 °C. The increase of metallising temperature to 1300 °C has almost no effect on the increase of the thickness of the coating ($h \cong$ 1 μm).[140]

The results of x-ray diffraction analysis indicate that the produced coating in the entire metallising temperature range consists almost completely of the higher tantalum carbide TaC. Evidently, this can be explained by high-intensity diffusion processes taking place at high temperatures. These processes determine the high rate of formation of the carbide phase at the interface. The lattice constant of tantalum carbide at different temperatures is unchanged and has the value 0.4456 nm.

The interaction of diamond with refractory metals has also been examined in Ref. 152. This was carried out using the methods of plasma spraying of metals on diamond heated to various temperatures from 1000 to 1500 °C. The results show that there is no chemical interaction of diamond with molybdenum, tungsten, and niobium during their plasma spraying on the diamond surface heated to 800 °C. When the surface is heated to 1000 °C and higher a thin carbide layer is observed, i.e. the rate of carbide formation increases with increasing temperature to which the diamond is heated.

The solid-phase interaction of diamond and metals of group VIII was investigated on an example of spraying nickel and iron on diamond at 1350±20 °C.[153] The results indicate that in interaction with the flux of atoms of these metals etching and graphitisation of diamond start at room temperature. Contact etching is preceded by a reaction the formation of intermediate metal-carbon complexes. For example, the diamond–iron system in the solid-phase is characterised by the active interaction with the formation of a diffusion zone of carbon atoms in the metal.

The experimental data have confirmed the increase of the intensity of diffusion of carbon with increasing temperature and contact time.

This is consistent with the published data on the coefficient of carbon diffusion in iron:

$$m = kc_0\sqrt{D\tau}$$

where m is the mass of dissolving diamond; k is a constant; c_0 is the concentration of carbon on the iron surface; D is the coefficient of diffusion of carbon in iron; τ is time.

The authors of Ref. 154 examined the interaction of diamond with nickel, cobalt and chromium. Cylinders with a diameter of 0.01 m were pressed from charges of metallic powders and this was followed by pouring in a diamond powder and then by pressing such a 'sandwich'. The specimens were heated in a vacuum furnace in dried hydrogen to 800-1400 °C.

The contact zone was examined on cross-sections by metallographic, x diffraction and local spectrum analysis. The results show that the contact the zone of Co with diamond as temperatures higher than 1100 °C is characterised by the presence of the zone of the solid solution of carbon in cobalt and by graphite precipitates and the grain bandits. Similar results were obtained in diffusion of carbon from diamond into nickel. However, as a result of the higher mobility of carbon in nickel in comparison with cobalt, the diffusion zone in this case was larger. After heating to 1400 °C, the structure of the surface layer consisted of a solid solution of carbon in nickel with graphite precipitates and graphite eutectic at the grain boundaries.

Contact of the chromium specimens with diamond in heating to 1100 - 1200 °C leads to the formation of a carbide layer, and in holding for two hours a layer consisting of a single zone formed; after 5 hr two zones formed, and after ten hours the carbide layer was clearly divided into three carbide zones. However, the absence of separation the carbide layer into individual layers does not yet mean that this layer is homogeneous. X-ray diffraction revealed the presence of carbides Cr_7C_3 and $Cr_{23}C_6$ in this layer.

Chapter 2

STRENGTH OF INTERFACIAL CONTACT OF DISSIMILAR SOLIDS

2.1 THEORETICAL CONSIDERATIONS

The strength of the interface between two different solids is determined (as in the case of the strength of a homogeneous solid) by the strength of the interatomic bond of the boundary atoms of each phase and by a set of the structure factors in the interfacial region.

It is possible to determine the relationship of the thermodynamic interfacial adhesion (this quantity can be regarded as approximately equal of the work of adhesion in the system, if one of the phases is transferred to the liquid state) and the real mechanical strength of the contact.

For example, if we consider the idealised case of an equilibrium contact system, where the completely smooth surface of one solid is connected with the surface of the other solid with defect-free and dislocation-free structures, and if the differences in the moments of separation of the atoms of one solid from the surface of the other solid are ignored, together with the residual stresses in the contact zone of different origin, then the work of adhesion in the reversible isothermal process of disruption of the contact is determined by the following expression:

$$W_A = \int_a^{\infty} F(x)dx = F(x)_{\max} x_k, \tag{2.1}$$

where $F(x)$ is the attraction force between the unit surface of the contacting phases at the given distance x of these surfaces; a is the equilibrium distance between the surfaces at their contact; x_k is the radius of action of interatomic forces. The force $F(x)$ increases initially with the change of the distance x, and then decreases to 0 (Fig.29). The maximum value of $F(x)_{\max}$ is the theoretical value of the rupture

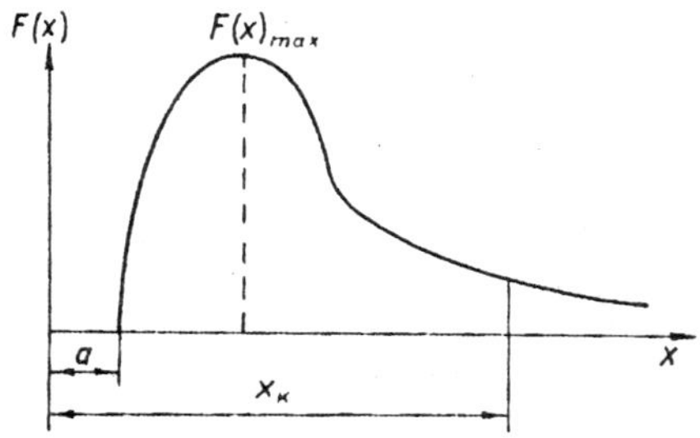

Fig.29 Diagram of the change of the the attraction force F of surfaces of the contacting phases on the distance between them.

strength of the contact.

In real cases, the fact that idealised conditions are not fulfilled - defects and dislocations in the structure, residual stresses, thermal stresses caused by the different of the coefficient of thermal expansion of the contacting materials, non-simultaneous dislocation rupture of the bonds, and the irreversibility of the process itself – lead to a disruption of the correlation between W_A and rupture strength. The measured value of the rupture stress is $\sigma_p \ll F(x)_{max}$. This ratio is identical with the ratio between the real and theoretical strength of the homogeneous solid. In the non-equilibrium contact systems the work of adhesion can depend on time and change as a result of the interaction processes (chemical reactions, dissolution, diffusion).

The problem of the relationship between the thermodynamic work of adhesion and the mechanical strength of the contact is discussed in the literature. There are different viewpoints on this subject.

In Ref. 155, the authors reported that, with other conditions being equal, the surface tension at the liquid-gas (σ_{liq}) interface has a direct effect on the bonding strength of the polymerised liquid with the solid. When the σ_{liq} decreases, the contact wetting angle in the system θ also decreases and the bonding strength increases. The authors concluded that, selecting the liquids with low values of σ_{liq}, it is possible to solve the problem of bonding strength.

At the same time, in the studies by Deryagin who also examined the adhesion of liquid and solid polymers to solids, it is claimed that the thermodynamic value of the bonding energy of the phases (the adhesion of the liquid polymer to the solid) is not linked by the correlation relationship with the mechanical strength of the contact.[156,157] Kobeko and Marei[158] did not detect even the qualitative relationship between the wetting of solids by water and the adhesion of ice to these solids.[159]

The following comments should be made regarding these results. In contact of solids with liquids with a low surface tension (20–70 mJ/m^2) the high degree of wetting indicates only that the liquid-solid adhesion is higher than the given values of surface tension, i.e. the very fact of wetting in the systems does not yet indicate a high absolute value of the work of adhesion. In the transition from non—wetting to wetting in these systems, the work of adhesion changes in a relatively narrow range (from 50 to 100 mJ/m^2).

The phenomenon of disruption of the solid–solid contact by separation of the polymer film (as in the experiments carried out by Deryagin and Krotova,[157]) is complicated as a result of the formation of a double electric layer at the contact and by some other reasons.

At moderate values of W_A and a small change of this volume, the effect of complicated factors, which change adhesion in comparison with its value in initial contact, the stresses at the interface, caused by different thermal expansion of the contacting substances (as in the studies carried out by P.P. Kobeko, et al.) are controlling: consequently, there is no direct relationship between W_A and the adhesion of solids.

In the systems containing liquids with a high surface energy, i.e. silicate melts ($\sigma_{liq} \cong$ 300–500 mJ/m^2) and, in particular, metallic melts ($\sigma_{liq} \cong$ 500–2000 mJ/m^2), the work of adhesion can change in a very wide range (from hundreds to several thousands of mJ/m^2). These high values of W_A in a wide range of the variation of this quantity should result in the presence of a mean-statistical correlation between the liquid-solid adhesion and the bonding strength of two solids.

Weiss[160] observed such a correlation in wetting platinum by molten glass of various composition, Fig. 30. O'Brien[161] also found the W_A-shear strength in wetting various solids by an enamel melt. The expansion coefficients of these solids were close to the expansion coefficient of the enamel (Table 6).

The adhesion phenomenon for diamond–metallic systems will be described in details in the following sections. Here, we shall examine the results of qualitative experiments. A copper droplet (diameter 5·10 m^{-3}), melted on the smooth surface of the diamond crystal, does

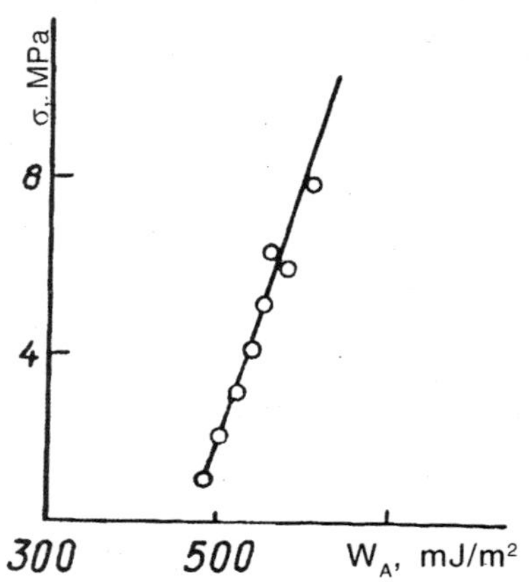

Fig.30 Correlation between the adhesion of a glass melt to platinum and the bonding strength of solidified glass.

Table 6 Wetting, adhesion and bonding strength of solidified enamel with different materials

Material of the solid phase	θ, °	W_A, mJ/m²	Shear strength of the enamel-material contact, kPa
Special material Ceramco	32	678	9800
Platinum	39	648	3800
Gold	59	563	3560
Graphite	141	82	0

not wet the diamond; the contact wetting angle is approximately 140°. The adhesion of liquid copper to diamond is low, approximately 200 mJ/m². After solidification of the metal during its cooling, the copper droplet separates from the diamond crystal almost without any force.

The addition to copper of small amounts of chromium (to 1–2%) leads to spreading and the contact angle decreases to 20°. The work of adhesion increases by an almost order of magnitude and reaches 1800–1900 mJ/m². The solidified droplet is strongly bonded to the surface of the diamond crystal; the separation force reaches thousands of newtons.

The authors of Ref.162 found that the mechanical strength of bonding of the sprayed metal films to glass is considerably higher in the system where chemical interaction takes place at the interface than in the system where only van-der-Walls forces operate. The bonding strength of the film of metals characterised by high affinity for oxygen (aluminium, chromium, etc) is considerably higher than that of metals with a low affinity for oxygen (tin, copper, silver, etc).

Thus, it may be assumed that the higher values of the thermodynamic quantity – the work of adhesion, the energy of the interfacial bond, are an essential although insufficient condition of the mechanically strong contact between the phases. The quantitative relationship of these quantities is the subject of special investigations and analysis.

2.2 METHODS OF DETERMINING CONTACT STRENGTH

The strength of interfacial contact can be determined by direct mechanical tests and also by indirect methods. In mechanical tests, measurements are taken of the fracture force of the contact and the area of the resultant surface (as when determining the strength of solids). Therefore, it is possible to measure the strength of contact in rupture, bending, shear, torsion. The resultant values of strength depend, as in the case of a homogeneous solid, on the dimensions of the specimens, their defectiveness and other factors.

It is difficult to determine the contact strength of dissimilar solids. There are cases in which fracture of the contact does not take place through the interface (adhesion rupture), and in many cases rupture can be of the adhesion–cohesion or even cohesion (through the body of one of the phases) nature. Therefore, when determining the strength of interfacial contact, the cohesion strength of the contacting phases must be higher than their adhesion.

Stresses concentrate at the interface if the properties of the contacting substances differ. They can form during the formation of contact - solidification or condensation of one of the phases, volume effects and non-uniform course of these processes through the entire bulk of the phase. Microstresses form because of the differences in the values of the lattice spacings of the contact phases and the coherence of the near-contact atomic layers, during chemical transformations in the interfacial area and the formation of new phases with the specific volume differing from that of the contacting phases, and also as a result of thermal effects and different values of the coefficients of thermal expansion of the contacting solids. The latter type of stresses can be calculated.[163]

The resultant stresses can be quite considerable and together with the applied forces or even on their own can cause fracture of the joint. There are methods of reducing or eliminating these stresses, for example, the use of a ductile material as one of the materials of the joint.

The mechanical methods of determining the contact strength are used in cases in which both contacting phases represent thick solid bodies. Ideal specimens for these tests are cylindrical or rectangular joints, as shown in Fig. 31. However, it is often difficult or even impossible to produce specimens of this shape, for example, in the case of the diamond crystal–metal system. In these cases, it is necessary to produce specimens contacting by the cylinder (cone)–plane system, Fig.32.

The contact strength of the coating with the substrate is determined by special methods. It is difficult to rupture such a joint because the thickness of the coatings is usually small (10^{-10}–10^{-4} m) and it is not possible to apply a force when clamping the specimens in, for example, the clamps of a tensile machine (as in the case of thick materials). In these cases, it is necessary to use additional materials and components of sufficiently large dimensions secured to the surface of the coating with the strength higher than the bonding strength of the coating

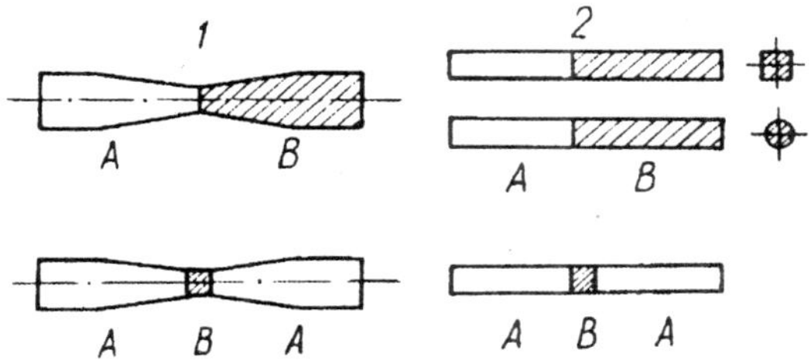

Fig.31 Specimens for tensile tests (1) and bend tests (2) by the two- or four-point method.

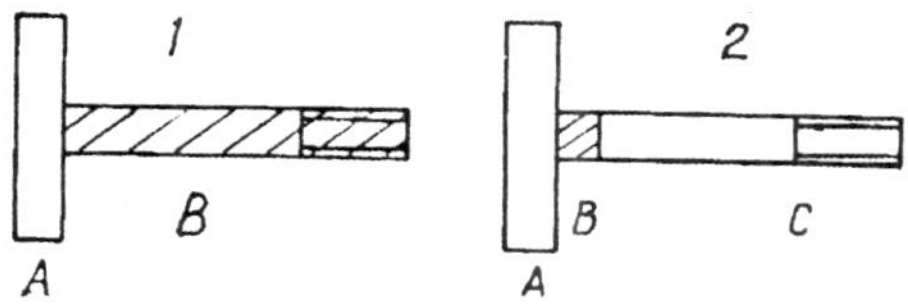

Fig.32 Joint–contact of the plane–cylinder type for tensile tests: *A* and *B* are contacting solid phases (1), *C* is the testing rod connected with the body *B* by the strength higher than the strength of contact *AB* (2).

to the substrate.

Here we shall examine methods of investigating the strength characteristics of joints of different types.

The main methods of determining the contact strength (including coatings) have been examined in Ref.157, 163–176. Adhesion bonding[164-169] or brazing[166, 170] of the testing rod or sheet, depending on the type of test in separation or shear, to the surface of the specimens with the deposited coating is used in mechanical tests of bonding strength of the coatings. In some cases, a flexible tape[171] is adhesion-bonded to the surface of the coating and is then separated together with the coating.

The authors of Ref.165, 166, 172–176 use the so-called pin method in which a pin is inserted into a hole in a sheet (the pin is made of the same material as the sheet), and a coating is deposited on the end of the pin situated on the same level with the surface of the sheet. This is followed by determining the separation force of the pin.

Relatively thick coatings can be tested for contact strength by separating them from the substrate using a special blade clamp.[176-179]

The scratching method has been developed – a loaded needle is pulled on the surface of the coating.[180] The needle, made of a hard material, has a rounded end with which it is pressed to the surface coated with film. During movement of the needle over the surface the load on the needle is gradually increased until it reaches the critical value at which the film separates and the needle leaves a clean trace - channel. The measure of adhesion in the film–substrate system is the maximum tangential load F applied to the indentor at the point of contact of the film and the indentor:

$$F = \frac{P\sqrt{W}}{\sqrt{r^2 \pi P - W}}, \tag{2.2}$$

where P is the Vickers hardness of the substrate; r is the radius of the indentor; W is the vertical load at which a continuous scratching line appears. This method has been used successfully for determining the bonding strength of the metallic coatings with glass.

There are special, mainly qualitative, methods of determining the bonding strength of the coatings with a substrate. The folding method[164,181] has been used for sheet materials with a coating in the cases in which the bonding strength is determined on the basis of the number of bends of the sheet to the first indication of separation of the coating.

In the method of cyclic impact loading[165,182] the surface of the coating is subjected to the multiple effect of a falling sphere or hammer. Evaluation is carried out on the basis of the nature of separation of the coating and also the development of separation of the coating in relation to the number of impacts.

The method of testing by extrusion[165] is based on the procedure in which a sphere is pushed into the sheet with the coating from the opposite side thus causing buckling of the coating and the substrate and also disruption of contact. After deformation of the specimens the stresses formed in the coating are examined. Some of the methods are based on using inertia and centrifugal forces. In the centrifugal method,[183] the coating is deposited on a cylindrical rotor and is then sectioned into strips parallel to the axis of the rotor to remove fracturing tangential forces in the coating. The rate required for separation of the film depends on the bonding forces of the film with the substrate.

The ultrasound vibration method[184] is based on exciting longitudinal oscillations of ultrasound frequency in a metallic cylinder whose end is coated. When the force formed in the coating under the effect of acceleration caused by vibrations (equal to the product of the square of frequency by the frequency of vibrations) exceeds the bonding forces on the interfacial surface, the coating is separated from the surface of the component. Measuring the frequency ω and the amplitude a of vibrations and knowing the dimensions of the interfacial surface S, the thickness of the coating δ and the density d of the coating material, it is possible to calculate the force causing separation of the coating from the equation

$$F = \omega^2 a\delta dS. \tag{2.3}$$

The normal separation method is regarded as most suitable of the examined methods for determining the interfacial contact strength (of thick solids or coatings).

It is important to ensure reproducibility of the results of strength tests of the brazed joints in separation or shear. For a number of brittle superhard materials, characterised by a large scatter of the experimental values of the contact strength, it is necessary to use a statistical approach to investigations – determination of the parameter characterising their strength properties and their scattering.

This is due to the fact that the presence in the body of polycrystals of superhard materials (for example, polycrystals of sphalerite or wurtzite modifications of boron nitride) of microdefects and heterogenities of the structure, with excessive stresses formed during testing and causing a fracture, has obviously the controlling effect on the strength characteristics of brazed contacts.

The statistical theory of brittle fracture, proposed by Weibull, and based on the model of the weakest member, determines the probability of fracture S at stresses σ_p

$$S = 1 - \exp\left(\frac{\sigma_p}{\sigma_0}\right)^m, \tag{2.4}$$

where m is the Weibull modulus, characterising the homogeneity of the material; σ_0 is the mean strength of the specimen which depends on its properties and dimensions.

This problem has been examined in details Ref. 185. The processing of the results of tensile testing cubis and wurtzite-like boron nitride steel joints using the Weibull theory carried out; the modulus values were obtained. These values were equal to 3.0 and 3.3, respectively. The results are presented in Fig. 64.

The low values of this parameter indicate the low degree of homogeneity of the polycrystals thus characterising the large scatter of the stresses of fracture of the joints.

2.3 MEASURING THE STRENGTH OF CONTACT OF DIAMOND WITH METALS

The strength of the diamond–metal joints was determined by the normal separation method.[122] A special device (Fig. 33) was developed for brazing the test rod to the surface of diamond with the deposited layer of the coating and for carrying out subsequent tests of the separation strength of the joint by the cylinder–plane method. Using guides, the device ensures strict axial centring of the rod in brazing and subsequent separation. Consequently, it is possible to remove almost completely the moment of forces and misalignment during fracture testing.

The diamond crystal 1 (Fig. 33) is secured into the steel holder 2 with a hole below the face. The edge of the hole contains bevelled areas for tight contact of diamond with the surface of the holder. At the bottom, the diamond is compressed by the pin 3, acting in the direction of the fracture force, thus preventing misalignment of the diamond during testing in the tensile machine and, consequently, preventing the formation of bending forces. The rod rests on the diamond face under the effect of the natural weight and the mass of the weight screwed onto the rod. The gap in the guides does not exceed $0.1 \cdot 10^{-3}$ m.

The thin sheet of the brazing alloy 7 is placed prior to brazing between the end of the rod and the face of the diamond. The device is then loaded into the furnace of vacuum equipment and held at the temperature determined by the melting point of the brazing alloy. The brazing alloy melts and connects the rod with the metallised surface of the diamond. After cooling and extraction from the furnace, the device is connected (without withdrawing the diamond crystal) to the clamps of

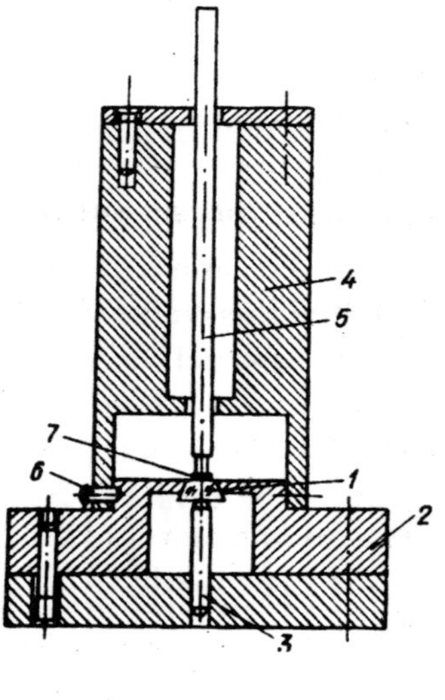

Fig.33 Diagram of the device for determining the strength of the interface between diamond and metal (alloy): 1) diamond; 2) holder; 3) moving pin; 4) vertical part with guides; 5) rod; 6) screw; 7) brazing alloy.

Fig.34 Device for determining the strength of contact between diamond and metal (alloy), connected to the clamps of a tensile testing machine.

the tensile machine and the rod with the coating are separated from the diamond crystal. During the tensile testing the device is in the turned-over position (see Fig. 34) to prevent the effect of the force of the mass of the device on the diamond–coating–brazing alloy–rod contact system. Flexible bars are used for connecting the device to the blocks of the tensile machine. The travel speed of the blocks is constant and equals approximately $1 \cdot 10^{-4}$ m/s. The separation force is recorded on the basis of the readings of the dynamometer and recorded diagrams.

The strength of the metallic film–diamond contact is determined as the ratio of the fracture force to the area of the contact spot clearly visible on the diamond face. The area of the spot is measured under a microscope.

Chapter 3

STRENGTH OF CONTACT IN THE DIAMOND-METAL SYSTEM

3.1 THE BONDING STRENGTH OF THE DIAMOND WITH METALS IN THE COMPACTED STATE

We shall examine the results of investigations of the mechanical strength of contact of diamond with metals and alloys brought together into contact in the liquid state.

The numerical values of the strength of the diamond–metal contact were determined for the first time probably in Ref. 2,44, 45, for the copper–silver–titanium, copper–chromium and copper–tin–titanium systems. The interface formed as a result of holding in vacuum at the required temperature during contact of the liquid alloy with the diamond surface.

The contact strength was tested by the normal separation method using the plane–rod set up. The rod was cast from the same alloy using a special graphite mould whose end contains a hole enabling contact of the melt with the diamond surface (Fig. 35). The results show that the alloy (copper, copper–silver), chemically inert to carbon, does not

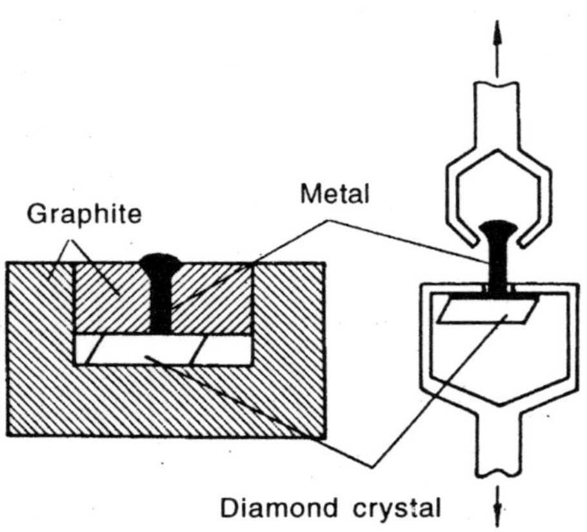

Fig.35 Diagram of melting the metal brazing alloy on a diamond crystal and testing the strength of the diamond–metal interface under tensile loading.

wet diamond in the liquid state and the strength of contact with diamond at room temperature (<1 MPa) is almost zero (or, in other words, the actual strength of contact is comparable with the thermal stresses formed in the contact region, as a result of differences in the coefficients of thermal expansion of the contacting diamond and metal).

The addition of titanium to the copper–silver alloy increases wetting and leads to an increase of the strength of bonding of the solidified alloy with the diamond surface. For example, for the alloy (Cu–72% Ag)–2% Ti, held in contact with diamond at 900 °C, the bonding strength with diamond after solidification of the alloy and cooling to room temperature is very high, 70 MPa, and at 1050 °C it is 10 MPa. For the (Cu–72% Ag)–2% Cr alloy, this parameter at 1000 °C is 60 MPa, and at 1060 °C it is 47 MPa.

On the whole, the results show a qualitative correlation of wetting and adhesion in the liquid state of the metal and diamond and the mechanical strength of the contact of the solidified alloy with the surface of the diamond crystal.

An important observation was made in the studies quoted above. The bonding strength increases with a decrease of the temperature of formation of the bond.

Superheating of liquid alloys that are in contact with diamond (copper-chromium) to 1300–1400°C resulted in the complete loss of contact strength. The contact properties of the copper–tin–titanium system (wetting, contact strength), in relation to diamond, have been examined in detail in Ref. 2, 45.

The authors of Ref. 2 examined the contact strength of the metallic alloy–diamond system using the method based on shearing the solidified melt droplet by applying a tangential force on the level of the equatorial diameter. The contact strength in shear was calculated from the equation:

$$\sigma = \frac{4L\tan\left(\theta - 90^\circ\right)}{\pi b^2}, \tag{3.1}$$

where σ is shear strength; L is load; b is the radius of the contact spot, θ is the contact angle.

In the examined case, the system is in a multiaxial stress state and the assumption, made by the authors regarding the uniform distribution of the shear stresses in the contact zone, is approximate. Consequently, the values of the strength of the diamond–metallic contact are often too high.

The data on the bonding strength of solidified alloys with diamond,

Table 7 Bonding strength of diamonds with C–20% Sn–2.1% Ti alloy in relation to contact temperature

t, °C	τ, min	θ, min	σ, MPa	Depth of erosion, μm
953 – 956	10	110 – 140	230	0.008
950 – 954	20	33 – 128	> 121	0.008
950	60	10 – 131	> 360	–
1002 – 1004	10	6 – 126	125	0.006
1000 – 1002	20	19 – 149	26 – 457	0.004
1000 – 1001	60	10 – 115	35 – 345	0.001 – 0.030
1050 – 1056	10	24 – 25	23	–
1051	20	10	42	–
1051 – 1055	60	5 – 12	18 – 72	0.025
1151	10	6	67	0.025
1156	20	6	3	–
1152	60	6	Very low	–

published in Ref. 2 and obtained in shear testing, are presented in Table 7. They show that the increase of temperature at which the interface between the diamond and the metal forms has a detrimental effect on contact strength.

The authors of Ref. 7, 46 found for a number of binary alloys that increasing the concentration of the active element, for example titanium, results in earlier formation of high mechanical contact strength (atlower concentration) in comparison with the moment of establishment of the maximum wetting of the diamond by alloys. For example, in Cu–Cr alloys wetting is observed at a chromium content higher than 0.08%. At the same time, the shear bonding strength is maximum (σ_{max}= 355 MPa) at a considerably lower Cr concentration (0.034%). The strength decreases with increasing Cr content.

In copper–titanium alloys, the maximum bonding strength in shear of 420 MPa was recorded at 0.032% titanium. High wetting was observed at a higher titanium concentration (12%).

Similar results were obtained for copper–vanadium alloys. High bonding strength, equalling 698 MPa in shear, was recorded for the

Table 8 Work of adhesion and bonding strength of a number of metals with diamond

Metal	T, K	W_A, mJ/m²	σ, MPa
Manganese	1518	2860	8
Germanium	1373	316	0
Tin	1073	260	0
Lead	1073	220	0

brazing alloy of the alloy containing 0.0048% V characterised by poor wetting of the diamond surface (θ ≅ 120°).

In addition to the reasons for not fulfilling the correlation between adhesion in the liquid and solid states, given in chapter 2, the following reasons can also be proposed.

Wetting depends strongly on the smallest contamination of the surface of the liquid alloy. This is especially evident in the case of alloys containing chemically active elements.

For example, in Ref. 7 it was shown that under especially clean experiment conditions, wetting of diamond and graphite by copper-titanium alloys already takes place at a titanium concentration of approximately 0.4–1.0%, in contrast to the data published in Ref. 46, where this value was 12%. Besides, the strength is a structure-sensitive property and depends on the concentration of the active element, carbide formation processes and special features of the structure of the interfacial region.

The authors of Ref. 8 examined the interaction and bonding strength of several metals (manganese, germanium, lead and tin) with the diamond surface. It was concluded that the high work of adhesion of the liquid metal to the diamond is accompanied by the relatively high bonding strength of the solidified metal with the diamond surface. The experimental results are presented in Table 8.

Similar investigations of the wetting of the diamond surface and the strength of the resultant contact were carried out in Ref. 47, 48, 186-188. For the alloys containing titanium, chromium, vanadium, niobium, the base was presented by a copper alloy with gallium (18.8% gallium). The alloy was selected because of its relatively high ductility and capacity of dissolving efficiently additions of many carbide-forming metals.

The tensile strength of the contact was determined in the device described in chapter 2.[122]

A tungsten testing rod was used; the brazing alloy which was used to braze the rod to the surface of the diamond crystal, was made of

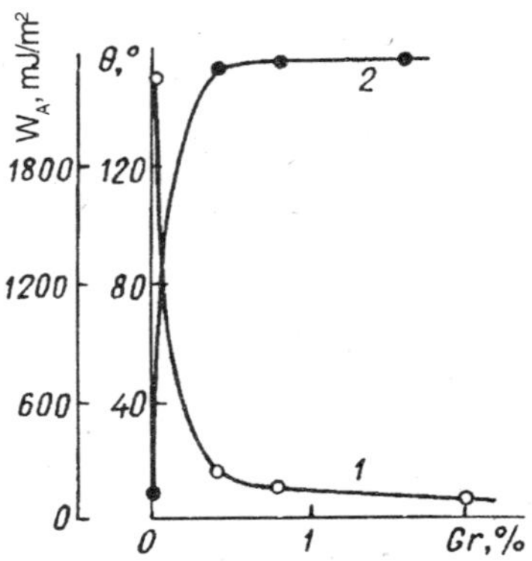

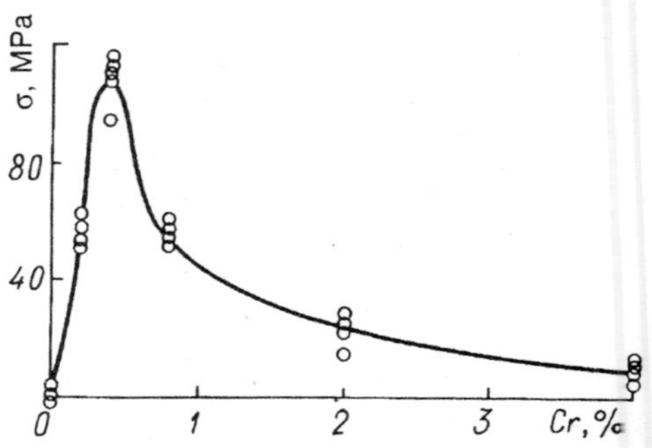

Fig.36 Dependence of the contact wetting angle (1) and work of adhesion (2) of Cu–Ga–Cr melts to the diamond surface on the chromium concentration at 1000°C.
Fig.37 Dependence of the bonding strength of Cu–Ga–Cr alloys with diamond on chromium concentration (holding time 15 min at 1000°C).

the tested alloy. At temperatures of 900–1000 °C, the intensity of interaction of this alloy with tungsten is low.

The diamond–copper–gallium–chromium alloy system

The experimental results show that the extent of wetting and bonding of the solidified Cu–18.8% Ga alloy with diamond is almost zero.

Small Cr additions (~ 0.4%) greatly decrease the value θ (from 150 to 48° at 960°C and to 18° at 1000 °C).

Cr additions of up to 0.4%, which greatly decrease the contact wetting angle, increase the work of adhesion to 1930 mJ/m^2 and resulted in sufficiently strong bonding (approximately 90–110 MPa) of the solidified droplet with diamond (Fig. 36, 37).

In this Cr concentration range there is a correlation between the work of adhesion and bonding strength. A further increase of the Cr concentration from 0.4 to 3.9% has only a slight effect on the content wetting angle and the work of adhesion. Bonding strength decreases to 25 MPa at a Cr concentration of the melt of 2.0%, and then decreases to 10 MPa at 3.9% chromium.[47] Figure 38 shows that the rapid increase of the work of adhesion of the liquid alloy to the diamond at the start of holding is accompanied by an increase of the mechanical bonding strength of the solidified alloy. Subsequently, with small changes of the adhesion of the liquid phase, the contact strength slightly decreases.

The values of the thickness of the carbide layer below copper–gallium–chromium alloys, held on the diamond surface for 15 minutes at 1000°C, are presented in Fig. 39.

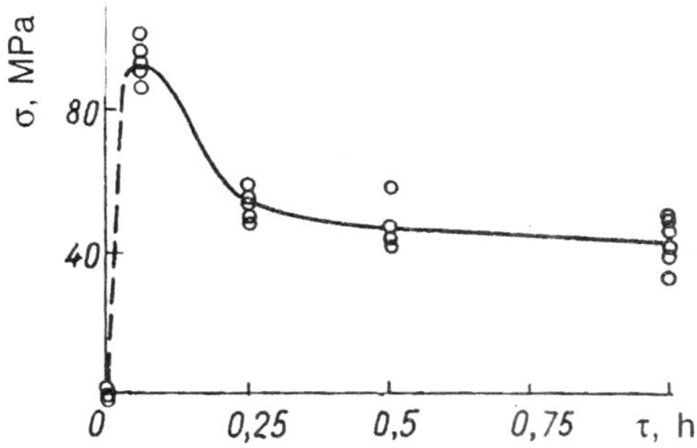

Fig.38 Dependence of the bonding strength of the (Cu–Ga)–0.8% Cr alloy with diamond on holding time at 1000°C.

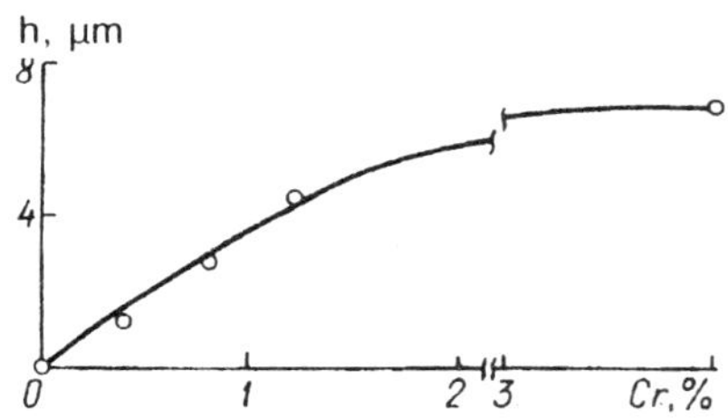

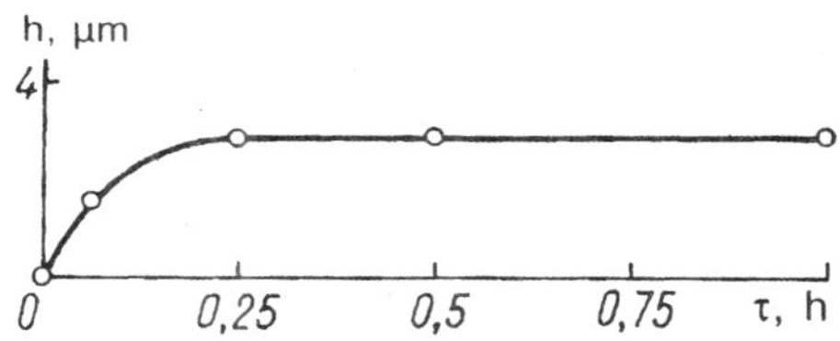

Fig.39 Dependence of the thickness of a carbide layer (*h*) at the diamond–Cu–Ga–Cr interface on chromium concentration (holding time 15 min at 1000°C).
Fig.40 (right) Dependence of the thickness of the carbide layer (h) at the diamond–(Cu–Ga) – 0.8% Cr interface on holding time at 1000°C.

With increasing Cr concentration of the alloy the thickness of the carbide layer the diamond increases and amounts to approximately 7.0 μm at a chromium content of 3.9%. The increase of the thickness of the carbide layer below the alloy at a lower chromium concentration (0.8% Cr) under isothermal holding at 1000 °C slows down. The maximum thickness of the layer, 3.0 μm, is obtained after holding for 15 minutes and then remains almost unchanged, Fig. 40.

The results of the investigations make it possible to draw several conclusions on the relationships governing the variation and relationship of wetting, work of adhesion, bonding strength of the solidified alloy with the diamond surface and the thickness of the carbide interlayer. The Cu–18.8% Ga alloy does not wet the diamond surface and poorly bonds with it because copper and gallium are chemically inert to carbon. High contact angles and the low values of the work of adhesion and bonding strength are determined by low physical (van-der-Walls) interaction forces. Greatly increasing the work of adhesion and mechanical bonding strength, the Cr addition of up to 0.4% forms a carbide interlayer as a result of chemical reaction. In this case, there is a correlation between the increase of the work of adhesion and the strength of the solidified droplet with the diamond, determined directly by separation. With a further increase of the Cr content in the melt the rate of increase of the work of adhesion decreases because of the saturation of the diamond–alloy interface by the adsorbed Cr atoms. At the same time, the thickness of the carbide interlayer increases from 1 to 7 μm as a result of polymolecular adsorption and diffusion processes. The smaller decrease of the bonding strength of the alloy (Cu-18.8% Ga)–0.8% Cr with increasing holding time in comparison with its decrease with increasing Cr concentration is caused by the fact that the carbide layer, after reaching a thickness of 3 μm in holding for 15 minutes, then remains almost constant during further contact of the melt with diamond.

The diamond–copper–gallium–titanium alloy system

The Cu–Ga alloys with different titanium additions were melted in an arc furnace with a cooled base. The liquid alloy was then poured into thick copper molds and after cooling extracted in the form of rods 7 mm in diameter, 100 mm long. The specimens produced by machining from these rods were subjected to the tensile test since these alloys are especially promising for diamond brazing. The data are in Fig. 43. The stresses at the diamond–metal interface were estimated using the approximate formula (see section 3.4.4, equation (3.6)) modified for the case of processing the metal alloys.[256]

The experimental results show that the addition of titanium to 0.74% greatly increases the degree of wetting (the contact wetting angle decreases from 160 to 19 degrees) and the work of adhesion of the melt to the diamond surface, Fig. 41.

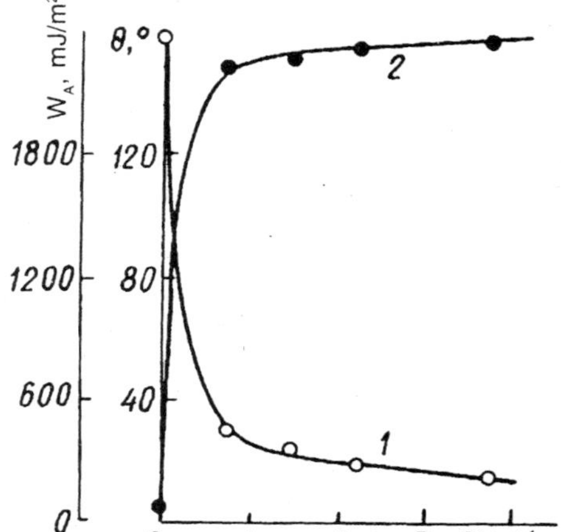

Fig.41 Dependence of the contact wetting angle (1) and work of adhesion (2) of Cu–Ga–Ti melts to the diamond surface on titanium concentration at 1000°C.

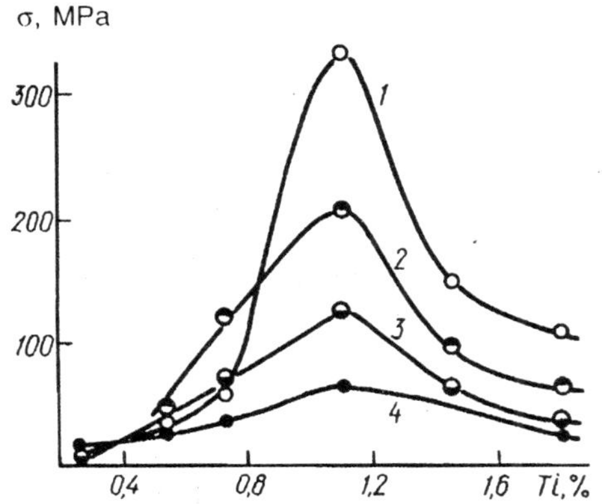

Fig.42 Dependence of the bonding strength of Cu–Ga–Ti alloys with diamond on the titanium concentration at a holding temperature of 1000°C and time: 1) 0.5; 2) 1; 3) 2; 4) 5 min.

It should be noted that the results obtained by the authors of Ref. 46 which show that the diamond surface is not wetted by the Cu–Ti alloy containing up to 12 of% Ti may be caused by the presence of difficult-to-remove oxides on the surface of the liquid alloy, as shown in Ref. 5 and 6. The addition of Sn to Cu and, in this case, Ga causes deoxidation of the alloy and makes it possible to obtain a high degree of wetting of diamond already at a low Ti concentration.

The addition of titanium to the copper–gallium base also increases the bonding strength of the alloy with the diamond, Fig. 42. The maximum strength of the diamond–metal contact, approximately 327 MPa, was recorded for the (Cu–Ga)–1.1% Ti alloy at a relatively short holding time, 0.5 min. The increase of the contact time of the melt with the diamond, like the increase of the of titanium concentration of the alloy to 1.5–1.8%, greatly decreases the contact strength.[187]

So, it can be concluded that the strength of the diamond/metal alloy contact in th ecase of alloys reacts chemically with carbon and is sensitive to brazing regime and alloy concentration.

Figure 43 shows the mechanical properties of the Cu–Ga–Ti alloys.

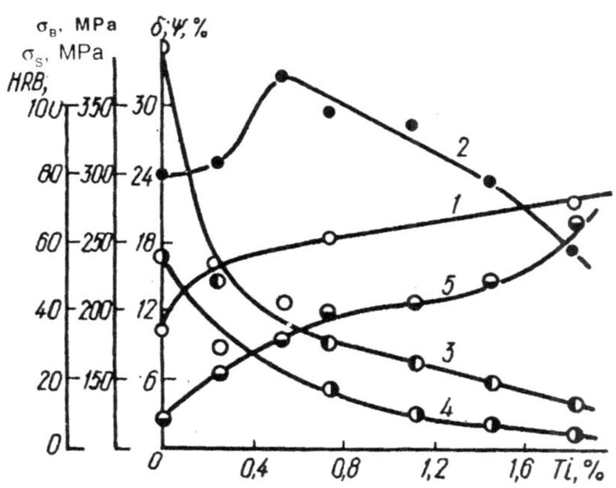

Fig.43 Dependence of the mechanical properties of Cu–Ga–Ti alloys on titanium concentration: 1) HRB hardness; 2) tensile strength σ_B; 3) relative elongation δ; 4) relative reduction area ψ; 5) thermal stresses at interface σ_S.

Table 9 Bonding strength with graphite of several adhesion-active alloys (brazing temperature 1000°C)

Composition of alloy, %	Holding time, min	σ, MPa
(Cu – 18.8 Ga) – 0.4 Cr	4	22 – 25
(Cu – 18.8 Ga) – 3.9 Cr	4	22 – 25
(Cu – 18.8 Ga) – 3.9 Cr	15	22 – 25
(Cu – 18.8 Ga) – 0.74 Ti	4	22 – 25
(Cu – 18.8 Ga) – 3.6 Ti	4	7 – 13
(Cu – 18.8 Ga) – 3.6 Ti	30	1 – 2

Comment. In all cases there was a large amount of graphite pulled out from below the melt droplet.

The maximum bonding strength of the graphite–Cu–Ga–Ti alloy contact in system is 22–25 MPa and is obtained, as in the case of diamond, at the minimum titanium concentration and holding time, Table 9. The increase of the titanium content and the contact time of the liquid droplet of the alloy with the diamond surface reduces the strength to 1–2 MPa.

The diamond–Cu–Ga–V alloy systems

The effect of vanadium on the bonding strength of the Cu–Ga alloy with diamond is similar to the effect of Cr and Ti.

Small vanadium additions (0.25%) result in a large increase of the degree of wetting and the work of adhesion (Fig. 44) and increase the strength of the diamond–metallic contact to 60 MPa. A further increase of the vanadium content of the alloy at the same work of adhesion decreases the bonding strength to 20 MPa (2.5% V). In the case of short-term holding of the melts with diamond (τ = 3 min), the bonding strength

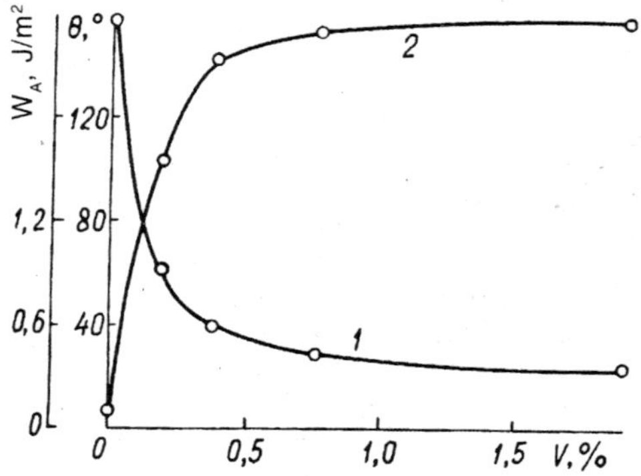

Fig.44 Dependence of the contact wetting angle (1) and work of adhesion (2) of Cu–Ga–V melts to the diamond surface on vanadium concentration at 1000°C.

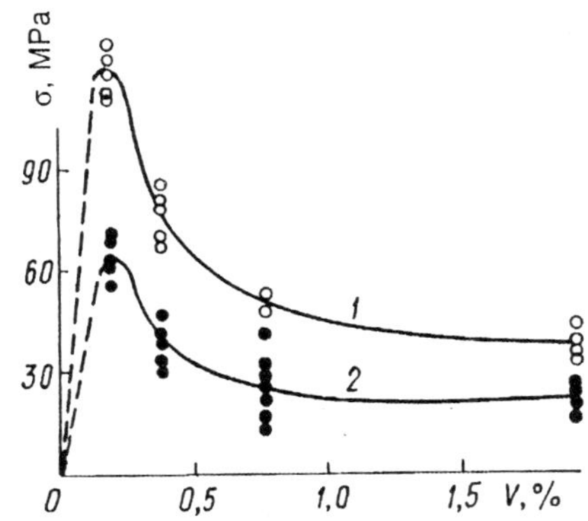

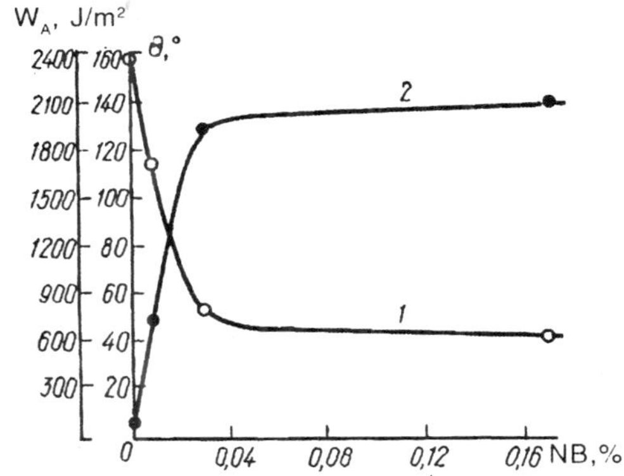

Fig.45 Dependence of bonding strength of Cu–Ga–V alloys with diamond on vanadium concentration at a holding time of 1000°C for 3 (1) and 15 min (2).
Fig.46 (right) Dependence of the contact wetting angle (1) and work of adhesion (2) of Cu–Ga–Nb melts to the diamond surface on niobium concentration at 1000°C.

is almost twice as high as that obtained after holding for 15 minutes (Fig. 45).

The diamond–Cu–Ga–Nb alloy system

The addition of niobium to the Cu–Ga alloy (to 0.17%) decreases the contact wetting angle of the diamond surface from 160 to 40°. The work of adhesion of the melt to diamond increases correspondingly from 0.1 to 2.1 J/m^2 (Fig. 46). According to the results of x-ray diffraction, microscopic and profile investigations, islands of the niobium carbide NbC forms in this case at the interface. When the niobium concentration of the alloy reaches 0.07%, the carbide layer becomes continuous, and at a niobium concentration of 0.17% its thickness increases to 1.0 μm.

The dependence of the contact strength of the Cu–Ga–Nb alloys with diamond on the niobium concentration is presented in Fig. 47. The contact strength of the alloys, containing up to 0.07% Nb, with the diamond is not high, evidently due to the discontinuous nature of the carbide layer. The maximum strength of the diamond–metallic contact (σ = 140 MPa) is obtained at a niobium content of 0.07% and a holding time of 1.0–5.0 min. The increase of the niobium concentration slightly decreases the contact strength (in contrast to the titanium addition, see Fig. 42).[188] Similar to examination of the diamond–Cu–Ga–Ti alloy system, to discuss the results, we present the mechanical properties of the Cu–Ga–Nb alloys in Fig. 48.

It should be noted that the smaller increase of the wetting, work of adhesion and the strength of the diamond–metallic contact when

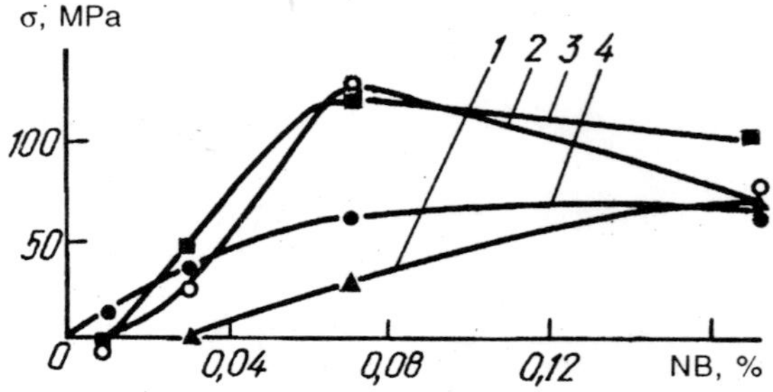

Fig.47 Dependence of the bonding strength of Cu–Ga–Nb alloys to diamond on niobium concentration at a holding temperature of 1000°C and time: 1) 0.5; 2) 1; 3) 5; 4) 10 min.

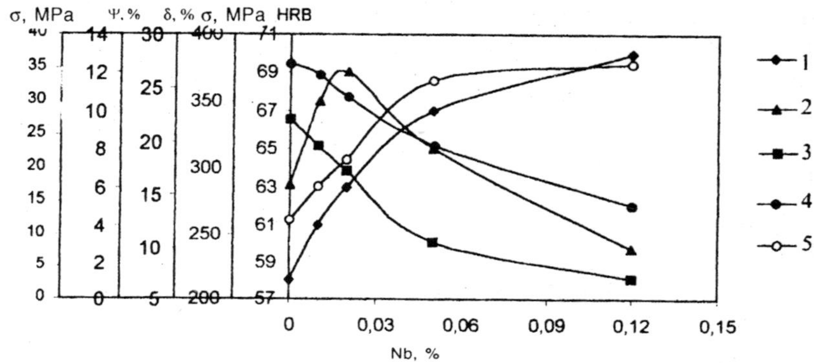

Fig.48 Dependence of mechanical properties of Cu–Ga–Nb alloys on niobium concentration: 1) HRB hardness; 2) tensile strength σ_B; 3) relative elongation δ; 4) relative reduction in area ψ; 5) thermal stresses at interface σ_n.

adding niobium to the Cu–Ga alloy in comparison with titanium is evidently explained by the lower chemical affinity of niobium for carbon and also by its lower solubility in the alloy.

Thus, the correlation between the work of adhesion of the liquid alloy and its bonding strength with diamond at room temperature exists only at low concentrations of chromium, titanium, vanadium and niobium in the alloys and a short holding time of the melts on the diamond surface. The increase of the concentration of the active addition or contact time leads, because of diffusion processes, to the formation of very thick carbide interlayers at the interface which decrease the strength characteristics of the joint between the metal and the diamond.

The experimental results show that in cases in which the effect of these processes is small, as observed in the case of the diamond–Cu-Ga–Ti alloy system and the diamond–Cu–Ga–Nb system, where thickness of the resultant carbide layer doesn't exceed 1 μm, the decrease of the contact strength with increasing amount of the active addition may be associated with a change of the mechanical properties of the alloy. For example, increasing niobium content increases the hardness and decreases the ductility of the alloy. The relative elongation δ decreases from 34.5 to 11.0%, and the relative reduction in area ψ from 17.1 to 6.9% (Fig. 48). This also relates to the strength of the alloy. The decrease of ductility with increasing niobium concentration is associated with the saturation of the soft Cu–Ga base with the hard intermetallic compound $NbGa_3$.[189] The corresponding calculations also show an increase of the Young's modulus of the alloy E and thermal stresses σ_x at the interface between the diamond and the alloy, causing the decrease of the strength of the brazed contact with the metal.

An even larger decrease of the strength of the diamond–metallic contact was observed when increasing the titanium concentration of the copper–gallium alloy (Fig. 42 and 43). When the titanium content was increased to 1.8%, the ductility of the alloy rapidly decreased (the value δ decreased to 2.9%, and that of ψ to 0.7%). The calculation results show that this was accompanied by a large increase of the level of thermal stresses at the interface (σ_x from 7.3 to 64.8 MPa), with the largest increase observed at a titanium content of the alloy higher than 1.1% (Fig. 43). Evidently, this influences the decrease of the contact strength of these alloys with diamond. The observed decrease of the ductility of the alloy with increasing titanium content is explained, as in the case of adding niobium, by an increase of the amount of brittle intermetallic compounds $TiCu_4$, Ti_3Ga, etc.[189]

The strength of the diamondmetallic contact, produced by the solid-phase method (compressing the diamond crystal and the metal on a flat surface) was examined in Ref. 190.

A truncated diamond pyramid was pressed with its tip to the surface of the metal at different temperatures in a vacuum. The loading force was approximately 50 N, holding time 3 min. After holding, the load was removed and the specimens were separated at the same temperature. The force, required for separating the specimens and related to the applied load, characterises the measure of adhesion interaction and is referred to as the adhesion coefficient. The results of tests for copper and several copper alloys are presented in Fig. 49.

The adhesion coefficient increases with increasing experiment temperatures for alloys, including copper. However, for the alloys, the temperature of the start of adhesion is higher. The authors explained this by a tendency of the alloy for increasing temperature of the start of the processes associated with some rearrangement of the crystal lattice. It should be noted that the reproducibility of the experimental data is not good.

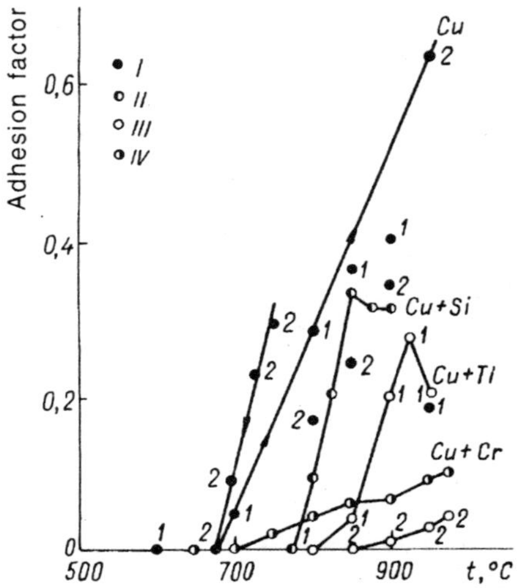

Fig.49 Dependence of the coefficient of adhesion to diamond of copper and its alloys on temperature: I) Cu; II) Cu–2.0% Si; III) Cu–0.45% Ti; IV) Cu–0.47% Cr (numbers 1 and 2 indicate the number of experiment).

The results of adhesion tests for other metals are presented in Fig. 50a. No specific conclusions can be drawn from the resultant dependences of the adhesion coefficient on temperature. Analysis of the results, presented in the adhesion coefficient–relative temperature coordinate system (T/T_m) shows that all tested pure metals are divided into two groups greatly differing in the temperature of the start of interaction (Fig. 50b).

The first group, characterised by the relatively low temperatures of the start of adhesion ($0.4–0.5T_m$), includes all metals characterised by active chemical interaction with carbon, including metals dissolving

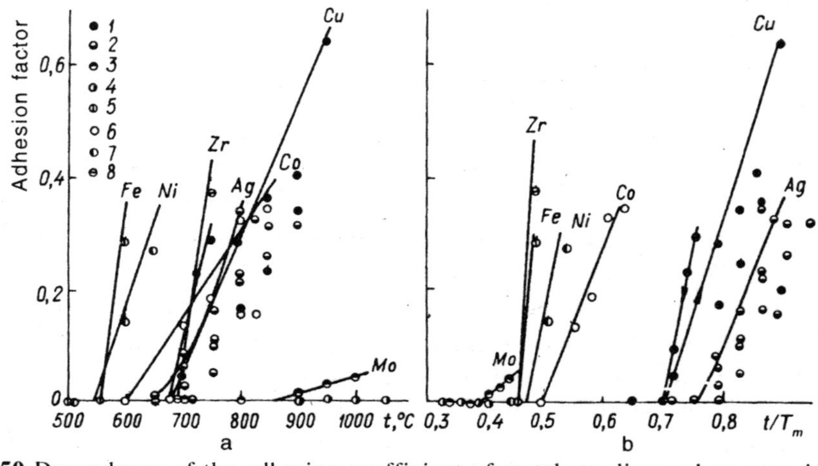

Fig.50 Dependence of the adhesion coefficient of metals to diamond on experiment temperature (a) and relative temperature (b): 1) Cu; 2) Ag; 3) Mo; 4) Ta; 5) Fe; 6) Co; 7) Ni; 8) Zr.

carbon (according to the classification presented in Ref. 6).

The second group includes the metals of the subgroup IB of the periodic table of elements (copper, silver) that are chemically inert to carbon. They are characterised by higher relative temperatures of the start of adhesion interaction (of the order of 0.7 T_m).

In interpreting these data it is necessary to consider, in addition to the chemical affinity of the metal for carbon, the ductility properties (yielding) of the metal (and the temperature dependence of these properties of the alloys), because this property controls the extent of real diamond–metal contact on the entire surface.

The numerical values of the contact strength were not determined in the work. The approximate estimate of this value is 30–70 MPa.

In Ref. 191, the authors determined the strength of retention of diamond in a copper–iron composite produced by infiltration with copper, in relation to test temperature. The values of the tensile strength of the specimens are presented below:

Temperature in heating, °C	20	200	400	600
Rupture strength of specimen, MPa	100	92	83	60

With increasing test temperature the strength of retention of the diamond in the binder decreases. In the majority of cases, fracture took place through the diamond grains, i.e. these strength values characterised the strength of the diamond itself. The interaction of the binder with the diamond is ensured by the presence of iron in the binder.

The author of Ref. 192 determined the strength of fastening diamonds (0.3 and 0.03 carat) metallised with nickel, cobalt, chromium and aluminium in depositing the metal initially by chemical reduction from the solution and then by the galvanic method, in cylindrical steel pins. The diamonds were placed at the very end of the pin to ensure that the material of the die surrounded the crystal from three sides. Gradually increasing force was applied to the part of the diamond projecting by 2/3. The qualitative data on the strength of securing the diamond crystal in the metallic die and the strength of the diamond–metal contact were obtained by testing the diamond grain embedded in the binder. The test results are presented in Table 10, which shows the values of the pull-out or fracturing force. The diamonds with the coatings are characterised by higher mechanical strength of fastening in the die.

Table 10 Separation force of grains (*N*) of diamond with different coatings

Diamond without coating	Ni-coating	Cr-coating	Al-coating	Co-coating
		0.3 carat		
250	1050*	240	–	–
350	550	500	–	–
240	350	460	–	–
290	500	460	–	–
250	–	400	–	–
		0.03 carat		
150	160	400*	550*	350
150	320	200	350*	200
100	350*	210	150	150
140	200	140	220	220
–	140	300*	310*	300*
–	180	350*	260	350*
–	–	160	280*	160
–	–	–	180	–

*diamond fracture

3.2 BONDING STRENGTH OF METALLIC COATINGS WITH DIAMOND

The authors of Ref. 133, 136–140, 193–196 published the results of systematic investigations of the strength of contact with diamond of the coatings of refractory carbide-forming metals — titanium, vanadium, niobium, tantalum, chromium, molybdenum and tungsten deposited by the contact-reaction method (by sintering the diamond in a mixture of a powder metallising agent). The coatings were deposited using the procedure described in Ref. 127 and 129. Attention was given to the effect of thickness, temperature (750–1200°C) and coating formation time (2–300 min).

Special features of the fracture of contact and the structure of the interfacial zone were examined. The bonding strength of the coatings was determined by the method described in the previous chapter, by normal separation of a test rod, brazed by its end to the flat surface of the diamond crystal. The dimensions of the natural diamond crystal were 1–1.5 carat, the size of the face $(5\times5)10^{-3}$ m; in most cases the face (111) was used. The brazing alloy was an alloy of copper with gallium (18.8% Ga, the maximum concentration of the solid solution) with the solidus temperature of 915°C, the liquidus temperature 970 °C. The rod was brazed in high vacuum at 1000 °C. The rod was made of molybdenum, its diameter was 3×10^{-3} m.

An important property of the alloy is its relatively high plasticity. This determines the relaxation of the stresses formed in the joint during cooling after brazing as a result of the difference of the thermal expansion coefficients.

The data for the thicknesses of the coatings of titanium, vanadium, niobium, tantalum, chromium, molybdenum and tungsten, obtained at different metallising temperatures and times, are presented in Fig. 51-53. When examining the temperature dependence of the thickness of coatings and also the bonding strength of the coatings with the diamond the metallising time was constant, 1 hour. This relatively long holding time is determined by the low rate of formation of the coatings.

With increasing metallising temperature the thickness of all examined coatings increases, and the thickness of the chromium coatings is greater than that of titanium, molybdenum and tungsten coatings, for

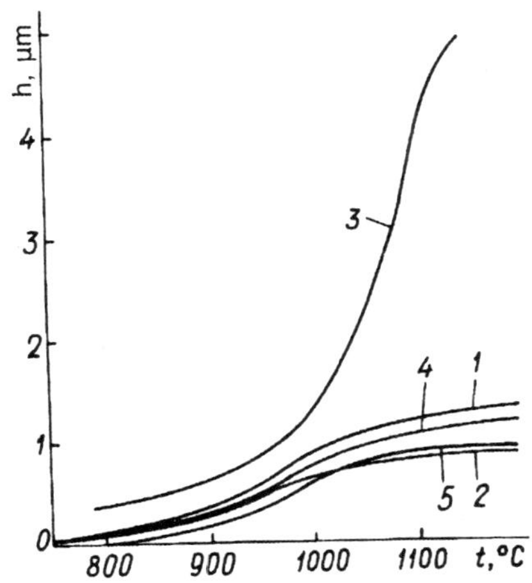

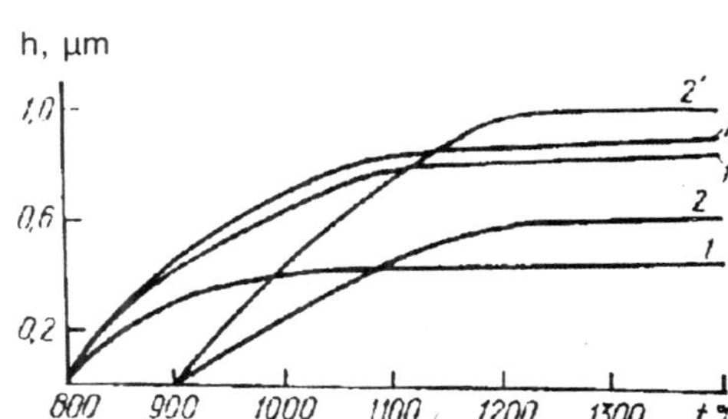

Fig.51 Dependence of the thickness of coatings on diamond on metallising temperature: 1) Ti; 2) V; 3) Cr; 4) Mo; 5) W.

Fig.52 Dependence of the thickness of niobium (1, 1', 1") and tantalum (2, 2') coatings on diamond on metallising temperature at the content of NaF activator: 1,2) 1; 1',2') 2; 1") 4%.

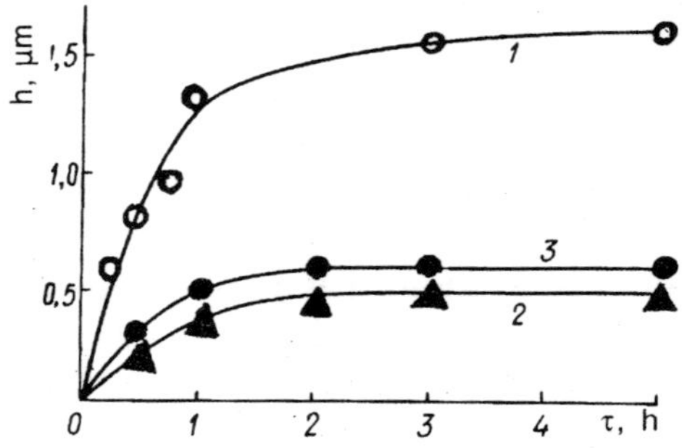

Fig.53 Dependence of the thickness of coatings on diamond on metallising time: 1) Cr (1000°C); 2) Ti (900°C); 3) Mo 950°C).

the same deposition conditions. For sample, at a metallising temperature of 1000 °C and a holding time of 1 hour, h_{Cr} = 1.3, h_{Ti} = 0.8, h_{Mo} = 0.7, h_W = 0.6 μm.

The bonding strength values were determined from 5–6 separate experiments. In cases in which the scatter of the data was large, the number of experiments was increased to 10–20. The measured results were processed by mathematical statistics methods.[197, 198]

The results of examination of the bonding strength of the titanium, vanadium, niobium, tantalum, chromium, molybdenum and tungsten coatings with the diamond surface are presented in Fig. 54–63.

The nature of the change of the bonding strength with diamond for each type of coating in relation to metallising temperature and time will now be examined.

The diamond–titanium coatings system

The bonding strength values of titanium coatings with the diamond surface are presented in Fig. 54.

At low metallising temperature (700–800°C) the bonding strength of the coating is low. The increase of temperature to 900°C increases the bonding strength to approximately 90 MPa. This value is maximum for the titanium coatings on diamond.

In the temperature range 900–1000°C the bonding strength decreases, almost to zero at 1100–1200°C.

To explain the effect of metallising time on the strength of the diamond–titanium contact, coatings were deposited at constant temperature ensuring maximum bonding strength with the diamond surface (900°C).

The titanium coatings deposited over a period of 15–20 minutes are very thin and poorly bonded to the diamond.

Metallising for 30 minutes increases the bonding strength to 54 MPa. The bonding strength of the titanium coatings increases when the metallising time is increased to 1 h, and then slightly decreases and

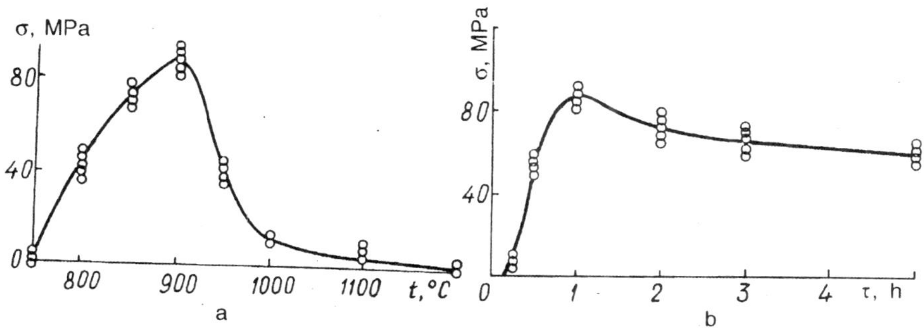

Fig.54 Dependence of the bonding strength of titanium coatings with diamond on metallising temperature at τ = 1 h (a) and on metallising time at t = 900°C (b).

equals approximately 2/3 σ_{max} after holding for five hours (60 MPa).

The diamond–vanadium coatings system

The dependence of the bonding strength of the vanadium coatings with diamond on metallising temperature and time is shown in Fig. 55.

The bonding strength with diamond of the thin coatings (h < 0.1 μm), produced at 800–850°C, is low. This can be explained by the fact that under the given metallising conditions the interaction of vanadium with the diamond surface is insufficient to produce a continuous carbide interlayer. Increasing temperature increases the bonding strength of vanadium with the diamond, reaching the maximum values of 100–110 MPa at 980°C. A further increase of temperature reduces the bonding strength. As in the case of the titanium coating at temperatures close to 1200°C, the bonding strength decreases to zero. With increasing metallising time the bonding strength of the vanadium coatings with the

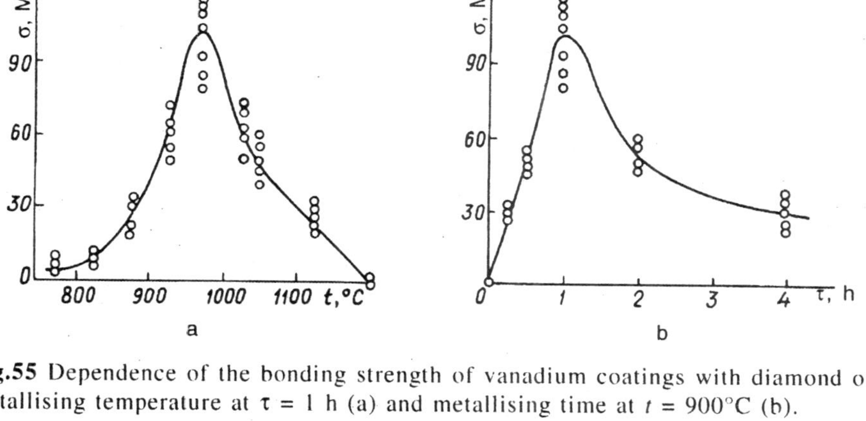

Fig.55 Dependence of the bonding strength of vanadium coatings with diamond on metallising temperature at τ = 1 h (a) and metallising time at t = 900°C (b).

diamond increases, reaching the maximum value after 60 min. A further increase of the metallising time reduces the contact strength but not so rapidly as in the case of the temperature dependence. After holding for 4 hr at 980°C, the contact strength is 1/3 of the maximum value (30 MPa).

The diamond–niobium coating system

Figure 56 shows the dependence of the bonding strength of the niobium coatings with diamond on metallising temperature (metallising time was 60 min) and the concentration of the activator.

The investigations were carried out in a wide temperature range (800–1400°C); this is explained by a small decrease in the contact strength of the niobium coatings with the diamond with increasing metallising temperature. The bonding strength of thin coatings, deposited at 800°C, is almost zero. The results of X-ray diffraction analysis indicates the absence of carbide formation at the interface, ensuring chemical bond between the coating and the diamond (Table 5).

Increasing metallising temperature increases the strength of the diamond–metal contact which reaches approximately 130 MPa (t = 950 °C, 3% NaF). A further increase of metallising temperature reduces the bonding strength. It should be noted that the maximum bonding strength of the niobium coating with the diamond is obtained when the metallising agent contains approximately 3% of the activator. The variation of the concentration of the activator to higher or lower values resulted in a large decrease of strength (σ = 50–70 MPa, the NaF content was 1, 2, 4%). Evidently, this can be explained by the optimum ratios of the phase in the coating at which no high stresses form. These stresses appear unavoidably at the interface as a result of the difference in the thermal expansion of the coating and the base.

The thin coatings (0.45 μm), deposited at an NaF concentration of 1% at 1100 °C, easily separate from the diamond surface (σ = 0) and a graphite interlayer is clearly visible below them. Thicker niobium coatings (NaF concentration exceeds 1%) retain the bonding strength

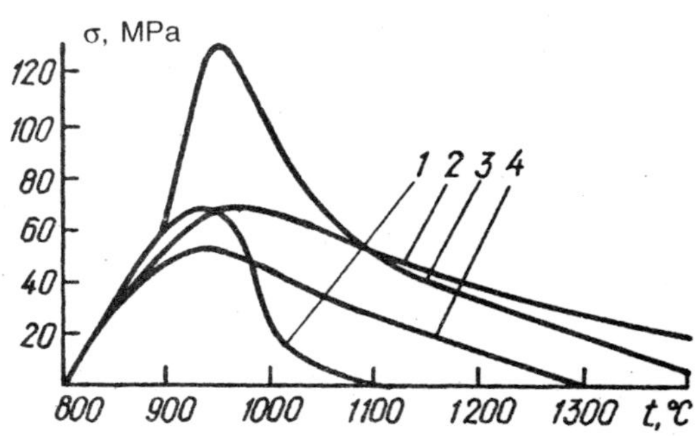

Fig.56 Dependence of the bonding strength of niobium coatings with diamond on metallising temperature (τ = 1 h) at the content of NaF activator: 1) 1; 2) 2; 3) 3; 4) 4%.

with the diamond surface different from zero even at higher temperatures of 1200–1300°C. This is due to the absence of graphitisation of the diamond surface.

Microscopy and x-ray investigations showed no traces of the graphite layer at the interface.

The diamond–tantalum coating system

The temperature dependence of the strength of the tantalum coating with the diamond at an activator concentration of 2 and 4% is shown in Fig. 57 (metallising time was 60 min). Increasing temperature increases the strength of the diamond–metal contact which reaches the maximum value of 105 MPa at 1100 °C (2% NaF) and then decreases.

Figure 58 shows the dependence of the contact strength on the NaF content of the metallising agent. It can be seen that the maximum strength of 120 MPa is obtained at a metallising temperature of 1100 °C and 2.5% NaF in the metallising agent.

The increase of the sodium fluoride concentration above 2.5%, and also the increase of the metallising temperature about 1100°C, greatly reduce the contact strength. Evidently, this is caused by high thermal stresses. This conclusion is supported by the fact that the coatings deposited on graphite at 1200 and 1300°C easily separate from its surface. In addition, in X-ray diffraction examination of the coatings, deposited on the diamond at 1300 and 1400°C, there were no traces of the graphite layer at the interface.

The diamond–molybdenum coating system

The experimental results show that the molybdenum coatings that start

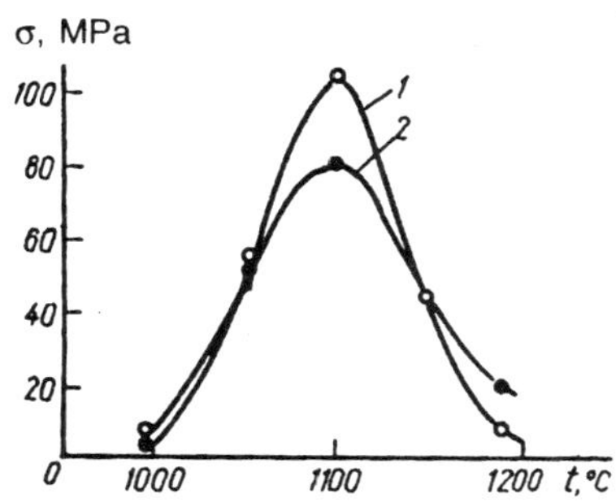

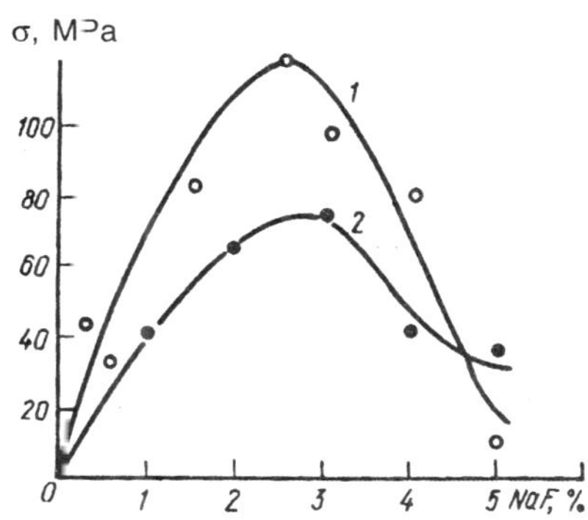

Fig.57 Dependence of the bonding strength of tantalum coatings with diamond on metallising temperature at $\tau = 1$ h and the content of NaF activator of 2 (1) and 4% (2)

Fig.58 Dependence of the bonding strength of tantalum coatings on NaF concentration in metallising agent at 1100 (1) and 1150°C (2).

to form at 800°C are poorly bonded to the diamond surface. Only a metallising temperature of 950°C with a holding time of 1 hour ensures the formation of molybdenum coatings strongly bonded to diamond (129 MPa). A further increase of temperature decreases the bonding strength, as in the case of the titanium coatings, almost to zero at 1200°C. The temperature nature of the dependence of the bonding strength of the molybdenum coatings with the diamond surface has a distinctive maximum at 950°C (Fig. 59a).

The dependence of the bonding strength of the molybdenum coatings with diamond on metallising time at a constant temperature of 950°C, optimum from the viewpoint of ensuring high strength, is shown in Fig. 59b. As in the case of the titanium coatings, the bonding strength increases with increasing metallising time to 1 hour, and this is followed by a decrease to 73 MPa after holding for 5 hr.

The diamond–tungsten coating system

The tungsten coatings which are the thinnest of all coatings examined here, at the same deposition conditions, also have the lowest bonding strength with the diamond surface. The experimental results are presented in Fig. 60. The maximum bonding strength was recorded for

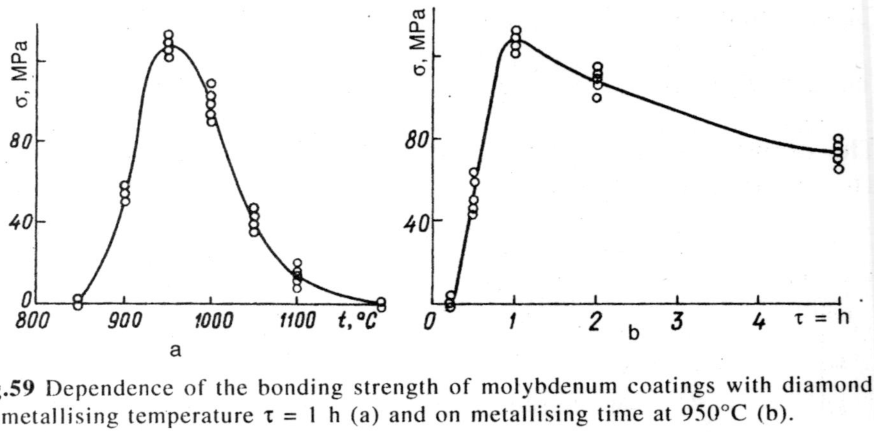

Fig.59 Dependence of the bonding strength of molybdenum coatings with diamond on metallising temperature τ = 1 h (a) and on metallising time at 950°C (b).

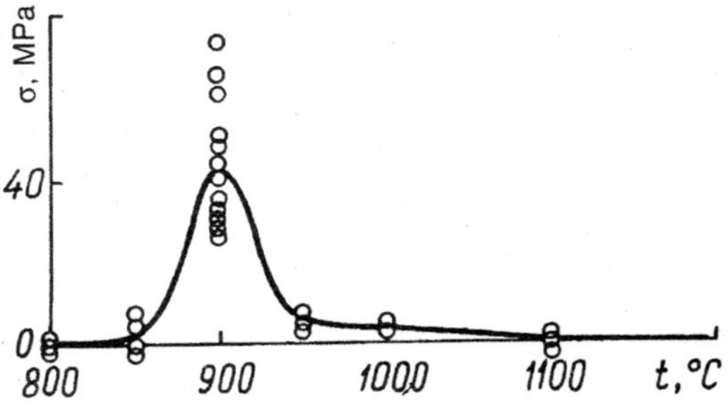

Fig.60 Dependence of the bonding strength of tungsten coatings with diamond on metallising temperature τ = 1 h.

the coatings deposited at 900 °C for 1 hr., and equals 44 MPa, with the largest scatter of the data (from 26 to 73 MPa). The decrease or increase of temperature results in a large decrease of the bonding strength of the tungsten coatings with the diamond surface. For example, the strength at 850 °C is 3 MPa, at 950 °C it is 5 MPa. In metallising at 1100 °C the tungsten coating is easily separated from the diamond. The bonding strength is close to zero.

The diamond–chromium coating system

The chromium coatings have been studied most extensively.[194, 195] This is caused by the fact that the range of thicknesses which can be produced for these coatings is very wide: from zero to tens of micrometers. It is therefore easier to describe the many factors affecting the bonding strength using the chromium coatings as an example. The temperature and time dependences of the bonding strength of the chromium coatings with a diamond surface are presented in Fig. 61 and 62.

At 900 °C and one hour, the chromium coatings produced have the bonding strength to diamond close to zero. As with any other examined types of coating, increasing metallising temperature at constant time also increases the strength of contact of diamond with chromium. The maximum strength was recorded at 1000 °C, 165 MPa (Fig. 61). A further increase of metallising temperature decreases the bonding strength. At 1200 °C the strength is 1 MPa.

To determine the effect of the temperature of formation of the diamond–chromium coating contact on strength without changing the thickness of the coating, the authors carried the following experiments. Chromium coatings were deposited on diamond crystals under the conditions ensuring the maximum bonding strength (1000 °C, 1 h). The specimens were then withdrawn from the metallising device and additionally annealed in a vacuum furnace under different conditions but without the metallising agent. After this annealing, separation tests were carried out. The nature of variation of the bonding strength in relation to the annealing temperature of the chromium coatings on diamond is shown in Fig. 61. When the annealing temperature is increased to 1150 °C the bonding strength of the coatings with the diamond remains almost constant, on the level of the maximum strength (165 MPa). Only annealing at 1200 °C results in a rapid decrease of the bonding strength, almost to zero. The dependence of the bonding strength of the chromium coatings with the diamond on the deposition time at temperatures of 1000 and 1200 °C is shown in Fig. 62.

At a metallising temperature of 1000 °C, holding for one hour results in the maximum bonding strength of 165 MPa, which then de-

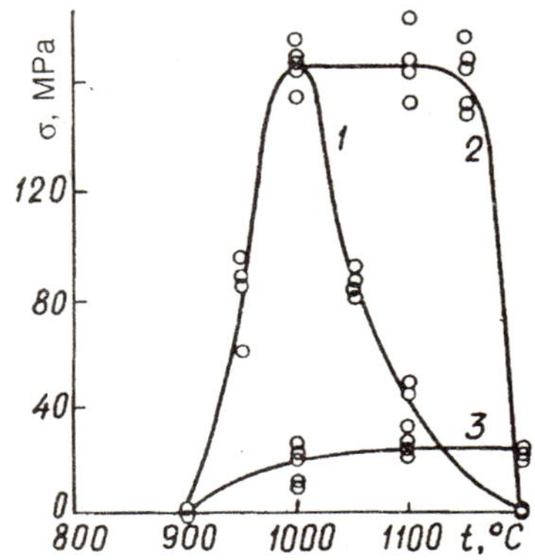

Fig.61 Dependence of the bonding strength of chromium coatings with diamond and graphite on processed temperature (τ = 1 h): 1) metallising (diamond); 2) annealing after metallising (diamond); 3) metallising and annealing after metallising (graphite).

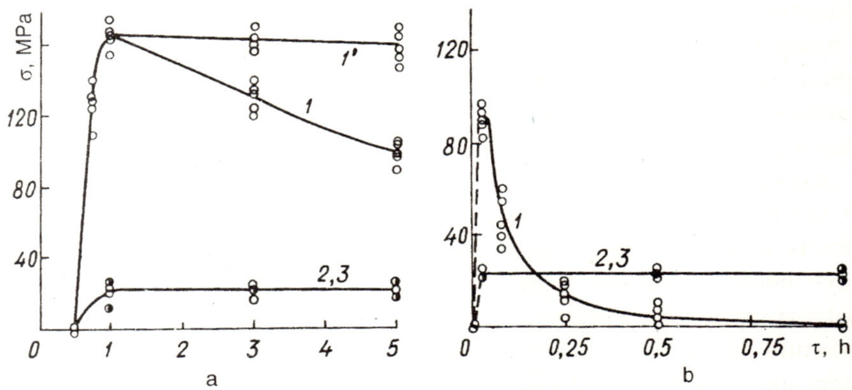

Fig.62 Dependence of the bonding strength of chromium coatings with diamond and graphite, deposited at 1000 (a) and 1200°C (b), on process time: 1) metallising (diamond); 1') annealing after metallising (diamond); 2) metallising (graphite); 3) annealing after metallising (graphite).

creases with increasing time (Fig. 62). 5 hour holding resulted in the formation of a chromium coating whose bonding strength with the diamond was approximately 100 MPa. Isothermal annealing (without the metallising agent) of the coatings, deposited under the optimum conditions, did not result in any decrease of their bonding strength with the diamond surface (Fig. 62a).

The time dependence of bonding strength for a metallising temperature of 1200 °C is characterised by the maximum of $\sigma \cong 90$ MPa, displaced to short holding times (2 min). With increasing metallising time the bonding strength rapidly decreases: at τ = 5 min the strength is 47 MPa, at 15 min it is 14 MPa, at τ = 30 min it is 4 MPa (see Fig. 62b).

The results of examination of the bonding strength of the chromium coatings with graphite are also presented in Fig. 61 and 62 (see curves 2 and 3).

The level of the bonding strength of chromium with graphite is lower than is diamond and is determined by the strength of graphite which is considerably lower than that of the diamond crystal. Therefore, separation took place through the graphite which was the weakest member in the rod–brazing alloy–coating–graphite contact system.

The bonding strength of the chromium coatings with graphite, equal to the strength of graphite (20–25 MPa), remained almost constant at metallising temperatures higher than 900 °C (τ = 1 h) or a holding time of more than 0.5 h (1000 °C). Metallising at higher temperatures of 1200 °C in a wide metallising time range ($0 < \tau \leq 5$ h) did not change the bonding strength. This is indicated by the absence of peaks and depressions of the strength curves, typical of the metallic coating–diamond systems. In addition to this, additional annealing of the previously formed chromium coating on graphite without the metallising agent had almost no effect on the bonding strength of the resultant contact pair. The summarised values of the bonding strength of the coatings with the diamond surface are presented in Fig. 63.

Recently, the results of separate investigations of the dependence of the bonding strength of the coatings of refractory carbide-forming metals with the surface of diamond and cubic boron nitride polycrystals have been published in the technical literature.[114, 152] The surface of polycrystals was deposited by plasma spraying, first with a layer of titanium 3–5 μm thick and then with copper. Titanium, forming the carbide, ensured the bonding of the coating with the specimens, whereas copper ensured wetting and spreading of the melt of the surface of the polycrystal during brazing. The specimens were heated during deposition of the titanium layer. The metallised specimens were brazed with AgCu brazing alloy to holders made of pure copper and steel. Brazing

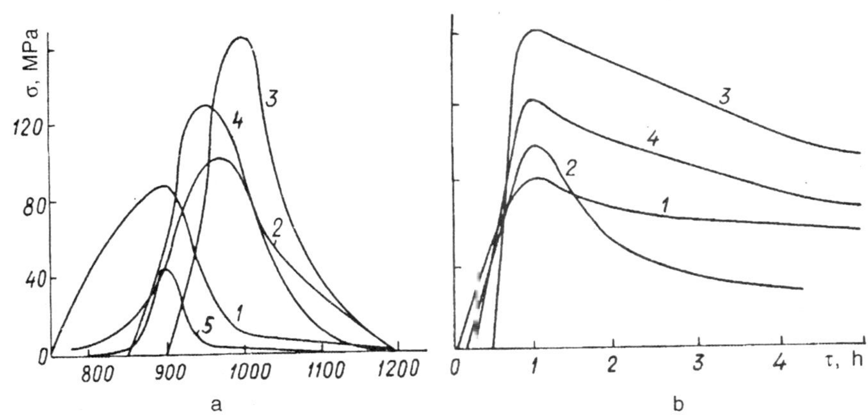

Fig.63 Dependence of the bonding strength of coatings with diamond on temperature at τ = 1 h (a) and metallising time (b): 1) Ti; 2) V; 3) Cr; 4) Mo; 5) W.

was carried out in a vacuum of (1–3)·10^{-3} Pa, holding time 3–5 min at the melting point of the brazing alloy. When the temperature to which the metallised specimens are heated is increased to 700 °C (metallising temperature), the strength of the brazed joints increases up to 50 MPa for diamond materials and to 70–80 MPa for cubic boron nitride. Further heating to 750 °C is characterised by a tendency to reduce the strength of the brazed joints. Higher strength values (~150 MPa for tensile and bend tests) were reached in Ref. 185 for boron nitride materials/steel joints, produced by direct brazing with CuAgTi and CuSnTi alloys (Fig. 64).

The experiments show[199] that the bonding strength of the diamond-titanium coating couple (in shear) is the function of the heat treatment time and is maximum after holding for 15–30 min (Fig. 65). This dependence is similar to that observed in Ref. 195.

The common shortcoming of the studies in Ref. 114, 152, 199 is the absence of investigations of the combined effect of temperature and heat treatment time on the bonding strength of diamond with titanium.

In Ref. 118 a molybdenum coating was deposited on diamond by cathodic sputtering in an inert medium. The diamond was heated to temperature not exceeding 800 °C. The thickness of the coating was varied from 0.058 to 0.2 μm. The bonding strength of diamond with molybdenum was investigated in a special device applying a torque to the rod brazed with copper. The bonding strength of the molybdenum film with the diamond surface in different experiments varied from 0.1 to 2.0 MPa. Evidently, the low bonding strength can be explained by the fact that the preheat temperature of the coating is insufficient to

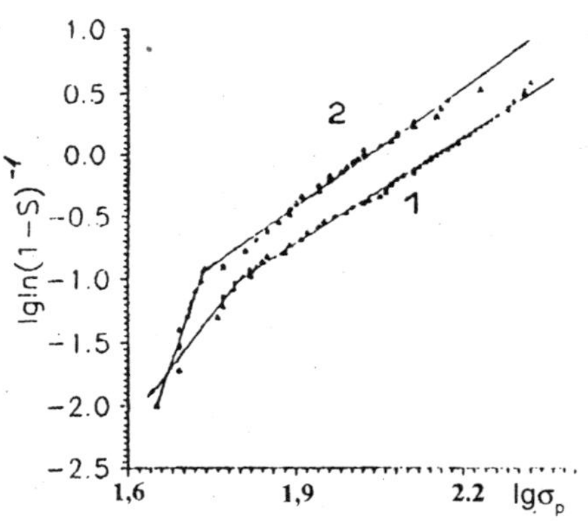

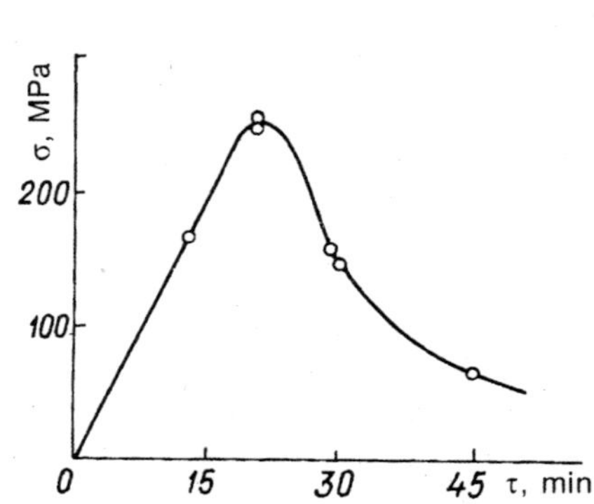

Fig.64 The Weibull plot for strength of brazed joints for superhard boron nitride materials/ steel assemblies. 1 – sphalerite form of BN, 2 – würtzite form of BN.
Fig.65 Dependence of bonding strength in shear of a titanium coating with diamond on treatment time (700°C).

produce a chemical bond of molybdenum with carbon. In addition, such a thin coating can dissolve in the copper brazing alloy when brazing the rod. This also reduces the bonding strength.

Low values of the bonding strength of the coating with the diamond surface were obtained in Ref. 113. Coatings of titanium, tantalum, chromium, tungsten and molybdenum were deposited by cathodic sputtering in vacuum. The thickness of the coatings on the diamond surface was approximately 1 μm. To stabilise the structure and reduce the internal stresses in the films and also improve their adhesion to the diamond, the specimens were subjected to short-term annealing at 900–1000 °C for 15 min. The highest bonding strength with the diamond surface was obtained in the case of titanium, tantalum, and tungsten coatings (Table 11), although the absolute values of the contact strength were low.

Table 11 Bonding strength of coatings with the diamond surface

Sprayed metal	Temperature of stabilising annealing of diamonds after metallising, °C	Separation stress, MPa
Titanium	900	12.8
Tantalum	1000	6.9
Tungsten	1000	11.8

3.3 BONDING STRENGTH OF COMPLICATED METALLIC COATINGS WITH DIAMOND

Metallising of diamonds with carbide-forming metals greatly increases the efficiency of tools as a result of increasing the strength of holding the diamond in the die. However, the common shortcoming of these coatings is their high affinity for oxygen which prevents diamond tools to be produced in oxygen-containing media: a layer of oxides not wetted by the brazing alloy forms on the coating surface.

Analysis of the literature data shows that the addition, to the composition of the coatings, of metals with a low affinity for oxygen, like copper, nickel or cobalt may have a positive effect on the wetting properties of the coatings.

Diamond–chromium–nickel, diamond–chromium–copper coating systems

Preliminary investigations were carried out into the kinetics of formation of chromium–nickel and chromium–copper coatings deposited on diamond by the method described in Ref. 200 from a powder metallising agent consisting of nickel or copper monoxide in vacuum heating a diamond-metallic mixture. Coatings of pure chromium were deposited for comparison. The chromium coatings started to form on diamond at temperatures around 800 °C. The monoxides of copper or nickel, added to the composition of the metallising agent, had almost no effect at this temperature. The thickness of the resultant coatings produced at 800 °C, metallising time 1 hour, does not exceed 0.04–0.1 μm. When the temperature is increased to 950–1000 °C, the thickness of the coatings increases, reaching 1–2 μm.[201]

Figure 66 and Table 12 show the dependences of the thickness of chromium–nickel and chromium–copper coatings on metallising temperature. When depositing the chromium–copper coating at temperatures higher than 1050 °C, the copper monoxide is reduced to pure Cu. This leads to the formation of a liquid phase and sintering of the diamond with the metallising agent.

At fixed temperatures, the thickness of the coatings continuously increases with increasing metallising time. The thickness of the chromium–nickel coatings is smaller and that of the chromium–copper coatings higher than that of the chromium coatings. This difference becomes larger with increasing content of the monoxides in the metallising agent.

The chromium coating is deposited on diamond in vacuum as a result

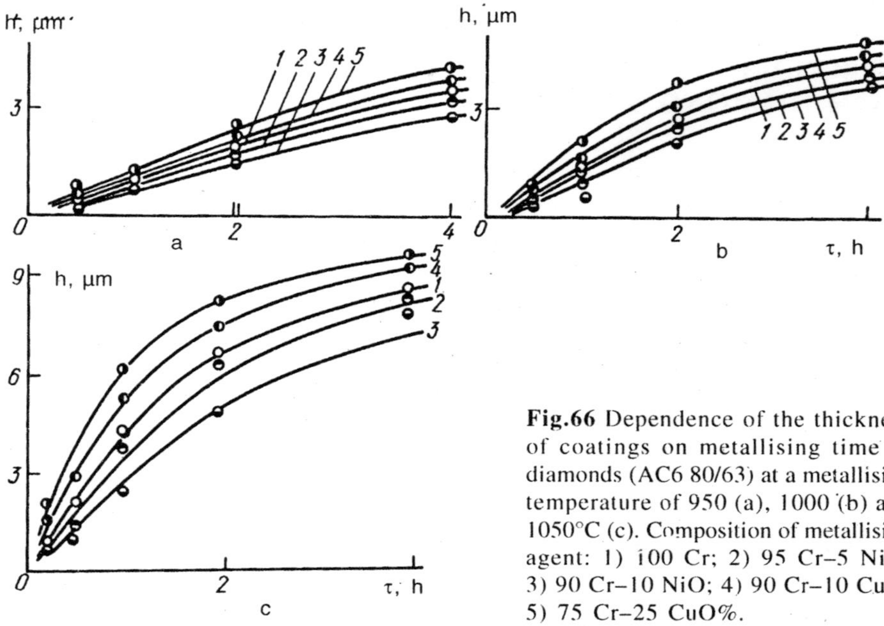

Fig.66 Dependence of the thickness of coatings on metallising time of diamonds (AC6 80/63) at a metallising temperature of 950 (a), 1000 (b) and 1050°C (c). Composition of metallising agent: 1) 100 Cr; 2) 95 Cr–5 NiO; 3) 90 Cr–10 NiO; 4) 90 Cr–10 CuO; 5) 75 Cr–25 CuO%.

Table 12 Thickness (μm) of chromium and chromium–copper coatings on diamond (metallising time 1 h)

O_2 content in metallising agent, %	Metallising temperature, °C				
	850	**900**	**950**	**1000**	**1050**
0*	0.3	0.4	0.7	1.3	4.2
10	0.4	0.6	0.8	1.6	5.2
25	0.5	0.7	1.1	2.3	6.0
50	0.7	0.8	1.4	3.0	–
75	0.9	1.0	1.8	4.1	–

*pure Cr coating

of the high pressure of chromium vapours resulting in the evaporation of chromium. The pressure of the vapours of chromium oxides is lower than that of chromium,[145, 202] and, consequently, during oxidation of chromium the thickness of the coatings formed on the diamond should decrease. This is confirmed by experiments. A chromium powder was oxidised in a muffle furnace in air at 600–700 °C for various periods of time. The degree of oxidation (oxygen content) was evaluated on the basis of the increase of the mass of the chromium powder. These oxidised chromium powders were used as a metallising agent when depositing coatings on diamond (Fig. 67). It can be seen that the thickness of the chromium coating decreases with increasing oxygen content in the metallising agent (curve 1). Similar thickness values were also recorded for chromium–nickel coatings (Fig. 67, curve 2), if the content of the nickel monoxide in the metallising agent is converted to the oxygen content in the agent.

The small decrease of the thickness of the chromium–nickel coat-

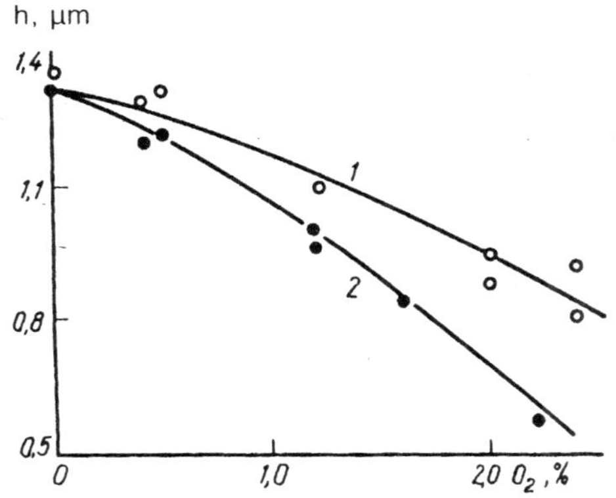

Fig.67 Dependence of the thickness of coatings on diamond powder 80/63 μm on oxygen content in the metallising agent (t = 1000 °C, τ = 1 h). Cr (1) and Cr – NiO (2).

ings, formed on the diamond, in comparison with the chromium coatings at the same oxygen content in the metallising agent is explained by the 'dilution' of the metallising agent with nickel which has a lower vapour pressure than chromium and its oxides.

Thus, when depositing chromium–nickel coatings on diamond, the nickel monoxide is reduced by chromium thus reducing the thickness of the coatings produced on the diamond. This is also confirmed by the results of x-ray diffraction analysis: nickel and chromium oxides were found in the metallising agent used.

When depositing a chromium–copper coating on the diamond, the thickness of the resultant coatings increases with increasing content of the copper monoxide in the metallising agent. As in the depositing the chromium–nickel coatings, chromium is oxidised during reduction of the copper monoxide. However, the thickness of the coatings does not decrease but, on the contrary, increases. This can be caused by the vapour pressure of copper and its oxides which is higher than the chromium vapour pressure at the metallising temperatures. In any case, the thickness of chromium–copper coatings is greater than that of chromium alone. The copper content in the coatings is higher than that in the metallising agent.[201]

When adding a small amount of nickel monoxide to the chromium metallising agent (5–10%) the minimum temperature at which a chemical bond forms between the coating and the diamond, ensuring adhesion strength of the bond, slightly decreases. For example, the strength of bonding of the coating, deposited from the metallising agent containing 95% chromium and 5% nickel monoxide at 900 °C, with diamond was 14 MPa, whereas for the purely chromium coating, deposited at the same temperature, it was equal to zero.

Increasing metallising temperature initially increases the strength of bonding of the chromium–nickel coatings with diamond. The strength reaches maximum and then decreases, Fig. 68. Increasing content of the nickel monoxide in the metallising agent displaces the maximum of the bonding strength of the coating to higher temperatures and the level of the maximum decreases (from 165 MPa in the absence of nickel monoxide in the chromium metallising agent to 24 MPa at 20% of nickel monoxide in the agent, at metallising temperatures of 1000 and 1100 °C, respectively).

The temperature dependences of the bonding strength of the chromium–copper coatings with diamond have also the form of curves with distinctive maxima (Fig. 69). The increase of bonding strength is determined by the formation of a carbide interlayer at the diamond–coating interface, i.e. by the formation of the Me–C chemical bond.

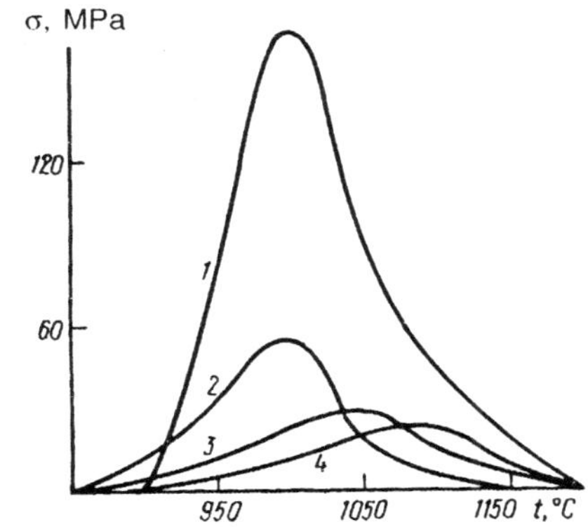

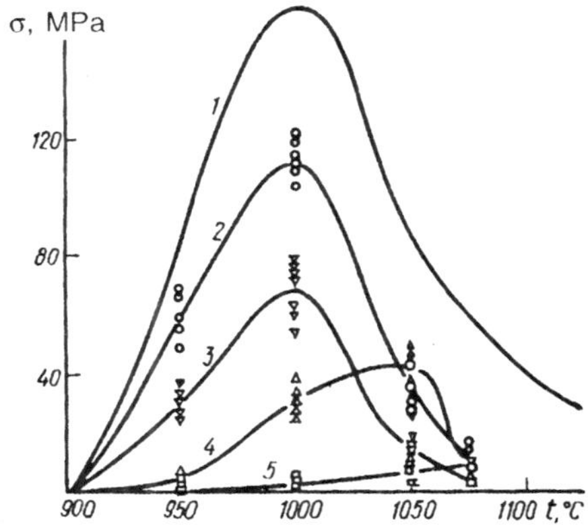

Fig.68 Dependence of the bonding strength with diamond of chromium–nickel coatings on metallising temperature (τ = 1 h) at NiO content of the metallising agent: 1) 0; 2) 5; 3) 10; 4) 20%.

Fig.69 Dependence of the bonding strength with diamond of chromium–copper coatings on metallising temperature (τ = 1 h) at a CuO content of the metallising agent: 1) 0; 2) 10; 3) 25; 4) 50; 5) 75%.

The increase of the content of Cu monoxide in the metallising agent decreases the bonding strength of the coatings with diamond. However, at a metallising temperature of 1000 °C and the CuO content of the metallising agent up to 20–25%, the strength remains quite high, 70–120 MPa. With increasing copper monoxide content the form of the σ–T dependence changes: the curves become flatter and the maximum is displaced to higher temperatures.[203]

These results can be explained as follows. With increasing content of the copper or nickel monoxide in the metallising agent the agent and, consequently, the coatings produced on diamond are 'diluted' by the component inert with respect to diamond. Consequently, the number of the bonds formed between the metal and carbon at the interface decreases and the rate of growth of the carbide layer is reduced and its integrity disrupted; this reduces the bonding strength.

When the metallising agent contains more than 50% copper monoxide, the maximum bonding strength is recorded at higher temperatures where the intensification of the carbide formation process results in the formation of a carbide interlayer ensuring sufficiently high bonding strength of the coating with the diamond, and the processes of softening of the diamond–metal contact are inhibited because of the presence of copper in the coating.

The diamond–molybdenum–nickel coating system

In contrast to the molybdenum coatings, which start to form at approximately 800 °C, molybdenum–nickel coatings start to form on a large scale at 650–700 °C. The process takes place as a result of complex interaction of molybdenum, nickel monoxide and carbon under the vacuum conditions. An important role is played by the processes of reduction of nickel monoxide by molybdenum with the formation of molybdenum oxides and also of the reduction of the latter with carbon with the formation of molybdenum carbide.[204]

Increasing metallising temperature and nickel monoxide content of the metallising agent increases the thickness of the coatings (Fig. 70a, b), which reaches 1.0–1.2 μm at the given metallising time.

The composition of the coatings in relation to the molybdenum and nickel monoxide content of the metallising agent, determined by chemical analysis, is presented in Table 13.

The dependence of the bonding strength of the molybdenum–nickel coatings with diamond on metallising temperature and the nickel monoxide content of the metallising agent is presented in Fig. 70c, d. Increasing metallising temperature (for all examined ratios of molybdenum and nickel monoxide content of the metallising agent) initially increases of the bonding strength of the coatings with diamond which reaches maximum values and then decreases.

At 10–15% of nickel in the coating high bonding strength (σ = 80÷100 MPa) is obtained at 750–800 °C. At higher temperatures, intensive carbide formation leads to disintegration and softening of the subsurface diamond layers as a result of diffusion of carbon atoms from diamond into the coating, thus causing softening of the diamond–metal contact.

At temperatures above 1000 °C, the decrease of the adhesion strength of contact of the coatings with diamond is also influenced by nickel as a catalyst of the process of graphitisation of diamond (this effect is especially strong at a high content of nickel monoxide in the metallising agent). For example, the bonding strength of the coating,

Table 13 Content of nickel (%) in molybdenum–nickel coatings deposited on diamond for metallising agents of different composition

NiO content in the metallising agent, %	Metallising temperature, °C		
	750	900	1000
9	9.9	6.0	3.4
25	28.5	16.9	13.7
50	58.4	35.1	31.0

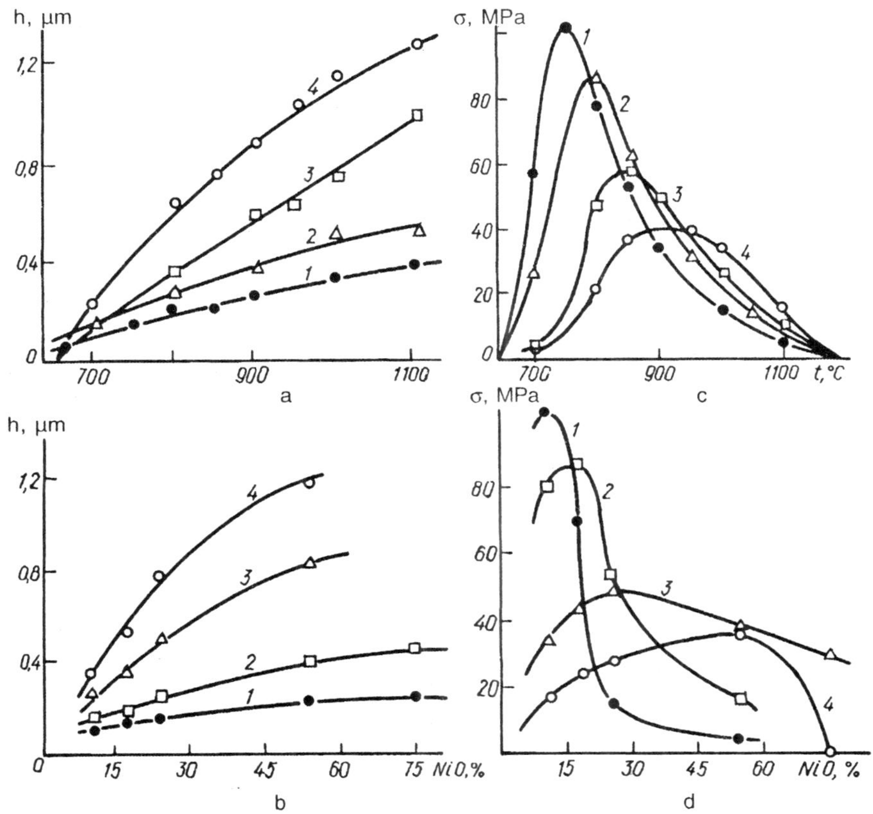

Fig.70 Dependence of thickness (a,b) and bonding strength (c,d) of molybdenum–nickel coatings on metallising temperature at different NiO content in the metallising agent (a,c) and NiO content in the metallising agent at different temperatures (b,d): a,c – 1) 9; 2) 17; 3) 25; 4) 50% NiO (τ = 1 h); b, d – 1) 750; 2) 800; 3) 900; 4) 1000°C.

deposited from the metallising agent, containing 75% nickel monoxide, under these conditions is zero (see Fig. 70d) because of rapid graphitisation of the diamond surface. This is clearly visible after separating the coating from the diamond – the contact spot is black.

Thus, the experimental results show that it is feasible to coat diamond with combined molybdenum–nickel coatings with high adhesion bonding strength at lower temperatures (750–800 °C). This improves the efficiency of retention of the initial strength properties of the diamond grains.

The diamond–chromium–molybdenum–nickel coating system

The chromium–molybdenum–nickel coating was deposited on diamond in vacuum $(2–3)\cdot10^{-3}$ Pa in a powder metallising agent, consisting of chromium, molybdenum and nickel monoxide, during heating and holding for one hour at a specific temperature. The composition of the coating was determined by chemical analysis.

The optimum number of experiments was determined by mathematical experiment design – a rotating plan of the second order. The experiments were carried out to obtain an empirical description of the dependence of the adhesion strength of content of the coatings with diamond which would make it possible to determine the strength of the effect on the contact of each of the examined factors (the molybdenum content of the metallising agent x_1, the nickel monoxide content x_2, metallising temperature x_3), and which would also be used as a basis for finding the optimum composition of the metallising agent and the temperature conditions of metallising from the viewpoint of the maximum contact strength.

Taking into account the apriori data on the process of deposition of chromium, molybdenum and molybdenum-nickel coatings on diamond, the authors selected the following levels and variation ranges of the factors (Table 14).

The experimental conditions and results are presented in Table 15. To increase the reliability of the results, three measurements of the contact strength of the coating with diamond were taken in each experiment.

The resultant mathematical model of the process (equation 3.2), whose regression coefficients were calculated using the procedure in Ref. 205, can be described as follows:

$$\sigma = 4.22 + 0.54x_1 + 1.93x_2 + 0.22x_3 + 2.24x_1x_2 - 1.80x_1x_3 - \\ -1.69x_2x_3 - 0.92x_1^2 + 0.91x_2^2 - 0.83x_3^2. \tag{3.2}$$

To evaluate the effect of each factor, select the composition of the metallising agent and the coating deposition temperature were used????, and the equation (3.2) is transformed:

$$\sigma = 4.44 + 0.54x_1 - 0.92x_1^2 \tag{3.3}$$

at x_2 = 0(25% NiO), x_3 = 0 (950 °C);

$$\sigma = 4.22 + 1.93x_2 + 0.91x_2^2 \tag{3.4}$$

Table 14 Coding, levels and ranges of variation of factors in determining the strength of the interface of Cr–Mo–Ni coatings with diamond

Variation range and levels of factors	Content in the metallising agent*, %		Metallising temperature, °C	Variation range and levels of factors	Content in the metallising agent*, %		Metallising temperature, °C
	Mo	NiO			Mo	NiO	
Zero level **x = 0**	40	25	950	Upper level x = + 1	55	40	1000
Variation range	15	15	50	Star points x = −1.682 x = + 1.682	15 65	0 50	865 1035
Lower level **x = −1**	25	10	900	Coded designation	x_1	x_2	x_3

*balance – chromium

Table 15 Conditions and results of experiments in determination of the strength of interface of Cr–Mo–Ni coatings with diamond

Experiment N°	x_1	x_2	x_3	σ', MPa	Experiment N°	x_1	x_2	x_3	σ', MPa
1	–	–	–	0	11	0	–1.682	0	36
2	+	–	–	2	12	0	+1.682	0	90
3	–	+	–	24	13	0	0	–1.682	0
4	+	+	–	128	14	0	0	+1.682	28
5	–	–	+	59	15	0	0	0	38
6	+	–	1	1	16	0	0	0	43
7	–	+	+	28	17	0	0	0	41
8	+	+	+	47	18	0	0	0	46
9	–1.682	0	0	9	19	0	0	0	39
10	+1.682	0	0	13	20	0	0	0	42

at $x_1 = 0$ (40 % Mo), $x_3 = 0$ (950 °C);

$$\sigma = 4.22 + 0.22x_3 - 0.83x_3^2 \qquad (3.5)$$

at $x_1 = 0$(40 % Mo), $x_2 = 0$ (25 % NiO).

As shown by analysis, at a low molybdenum and nickel monoxide content (to 30-40 and 10%, respectively), i.e. at a high chromium content in the metallising agent and low metallising temperatures (900–950 °C) no coatings form on the diamond and, consequently, the strength of the diamond–metallic contact is low. When the metallising temperature is increased to 1000–1050 °C, the thickness of the coating rapidly increases and the contact strength with diamond increases. A further increase of the temperature results in softening of the diamond-metallic contact because of the extensive formation and growth of the carbide at the diamond–coating interface, thus increasing the thermal stresses and resulting in the formation of a porous zone in the subsurface layers of diamond because of the difference in the diffusion coefficient of the metal into diamond and carbon into metal (see chapter 3.4).

When the molybdenum content in the metallising agent is higher than 40–50% and the nickel monoxide content exceeds 20–40%, a coating forms on the diamond already at a temperature of 750–800 °C. The process takes place as a result of complex interaction of molybdenum, nickel monoxide, chromium and carbon in vacuum. An important role is played by the processes of metallothermal production of nickel by molybdenum and chromium with the formation of their oxides and the reduction of the latter by carbon with the formation of carbides. The temperatures above 1000 °C decrease the strength of the diamond–metal contact also as a result of the formation of brittle carbides and porous zone in the subsurface diamond layer, as mentioned previously.

When the nickel monoxide content of the metallising agent is higher than 50–60%, the surface of the coating produced on the diamond is oxidised by the oxygen of the nickel monoxide. Consequently, the coating is not wetted by the brazing alloy.

Figure 71 shows the experimental dependences of the thickness, chemical composition and bonding strength of chromium–molybdenum–nickel coatings with diamond (the coatings were deposited at 900 and 1000 °C), on the composition of the metallising agent. It can be seen that at a metallising temperature of 900 °C the maximum value of the contact strength (120–130 MPa), are obtained when depositing the coating from the metallising agent containing 10–20% chromium, 45–55% molybdenum and 30–40% nickel monoxide. The coating with a thickness of 0.5–0.7 μm consists of 5–15% chromium, 40–55% molybdenum and 30–40% nickel. At a metallising temperature of 1000 °C there are two regions with high bonding strength: 1) when the metallising agent contains 5–50% chromium, 50–55% molybdenum and 35–40% nickel monoxide; 2) with high chromium content: 70–100% chromium, 0–30% molybdenum and 0–10% nickel monoxide.

These results make it possible to select the metallising agent and temperature for depositing chromium–molybdenum–nickel coatings of diamond to ensure the formation on the diamond of the coatings of the required composition with the given adhesion strength of the contact.

The diamond–chromium–molybdenum–copper coating system

As regards the chromium–molybdenum–copper coatings, it should be noted that their thickness is considerably greater than that of the chromium–molybdenum–nickel coatings at the same oxide content of the metallising agent. This is the result of the fact that, as already mentioned, the addition of the nickel oxide to the chromium metallising agent decreases the thickness of the chromium–nickel coating, formed on the diamond, as a result of oxidation of chromium during metallising. Although the addition of molybdenum to the metallising agent, consisting of chromium and nickel monoxide, slightly increases the thickness of the coating, this thickness is still smaller than that of the chromium coating.

In addition, the introduction of the copper monoxide to the chromium metallising agent increases the thickness of the chromium–copper coatings. The addition of molybdenum to such a metallising agent should

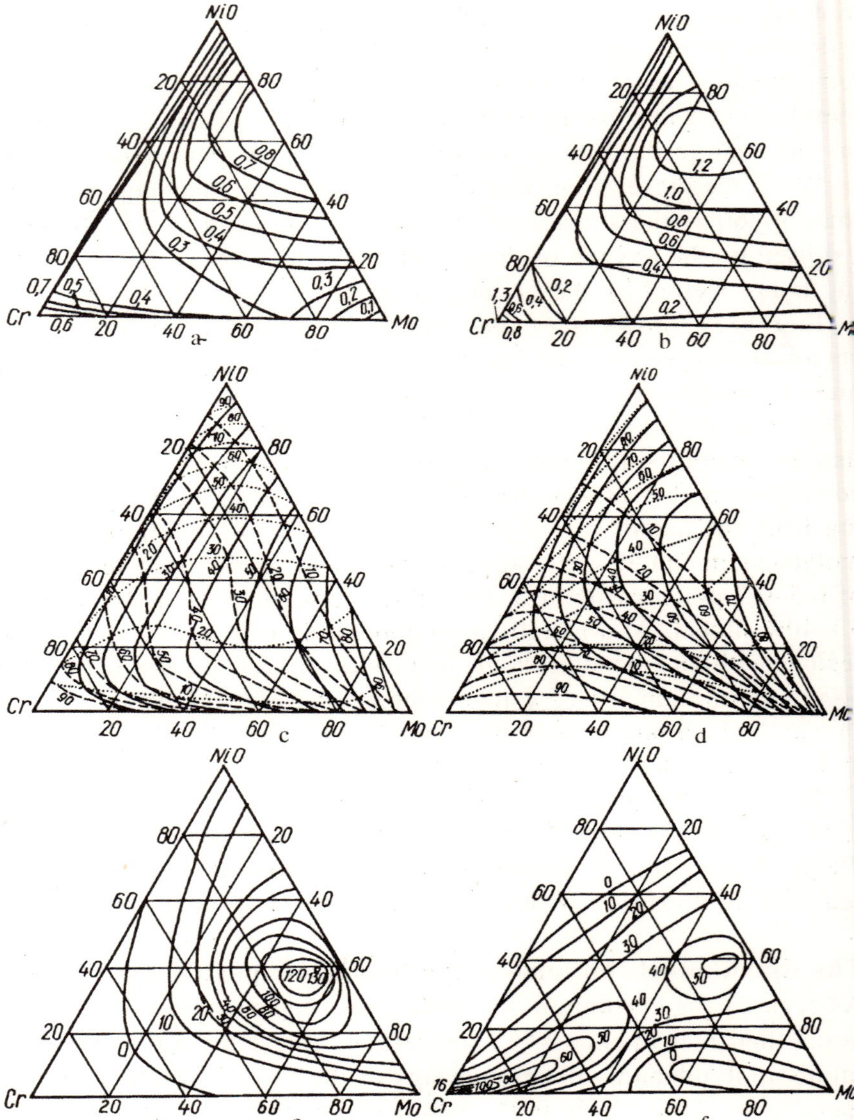

Fig.71 Dependence of the properties of chromium–molybdenum–nickel coatings deposited on diamond (τ = 1 h) at 900 (a, c, e) and 1000°C (b,d,f) on the composition of the metallising agent: a, b) thickness of coatings (μm); c,d) chemical composition of the coating (%); e, f) bonding strength (MPa). Solid lines – Mo, broken lines – Cr, dot-and dash-lines – Ni.

result in an even larger increase of the thickness of the chromium-molybdenum–copper coatings deposited on diamond. This is also observed in reality, Fig. 72. The strength of bonding of the chromium-molybdenum–copper coatings with the diamond depends on the ratio

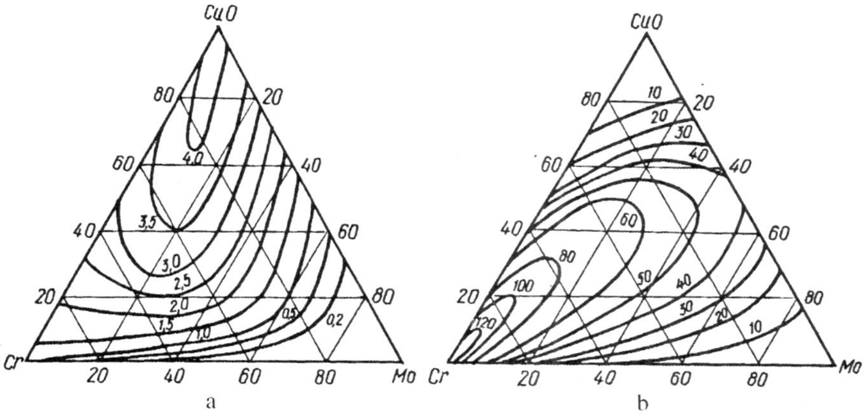

Fig.72 Dependence of the thickness (μm) (a) and bonding strength (MPa) (b) of chromium–molybdenum–copper coatings on diamond on the composition of the metallising agent at a metallising temperature of 1000°C (τ = 1 h).

of the components forming the coating. The bonding strength of these coating decreases with increasing chromium content of the metallising agent. However, when the molybdenum copper content in the coating is in the range 20–30%, the bonding strength is relatively high, 60–80 MPa. In comparison with the chromium–molybdenum–nickel coatings, the thickness and strength of bonding of the chromium–molybdenum-copper coatings with diamond are higher in a relatively wide range of the content of the components of the metallising agent, under the same temperature–time conditions of metallising.

The results of examination of the strength properties of the metallic carbide-forming simple and complex coatings on diamond make it possible to select the composition and the conditions of deposition of coatings on diamond that can be used in producing various types of diamond tools.

3.4 RELATIONSHIPS GOVERNING THE FORMATION OF STRONG CONTACT BETWEEN DIAMOND AND METAL

3.4.1 Nature of failure of the diamond–metal contact and the structure of the interface region

To explain the mechanisms of interfacial bonding, the structure and constitution of the contact region, investigations were carried out into the nature of separation of the metal from the diamond surface (at the interface, through the body of the diamond), the fracture surface, the microstructure and phase composition of the contact region. Measurements were taken of the depth of erosion of diamond after separation of the metallic solidified droplet from its surface.[2]

Detailed profile investigations of the interface and the relief in the zone fracture of contact were carried out in Ref. 206, 207. Profile measurements were taken in a profilometer–profilograph. This device makes it possible to determine the microrelief of the specimen during passage of a feeler needle on its surface due to the presence of indicating and writing units electrically connected with the sensor/needle (magnification of the device to 10^5 times).

The following contact systems were selected for profilographic measurements: diamond–chromium and diamond–copper–gallium–chromium. Investigations were carried out into the surface roughness of chromium coatings deposited at different temperatures and holding times, the microrelief of the diamond surface below the coating (after removal of the coating by etching, after fracture tests), the microrelief of the carbide layer formed on the diamond surface below the metallic alloy, after removing the alloy of copper and gallium by dissolution in special reagents. In addition, the thickness of the coating protruding above the diamond surface and 'grown' into it as a result of diffusion processes was determined.

In profilographic measurements, the so-called zero level was set on the diamond surface and the height of depressions or sections protruding above this level was measured. For this purpose, part of the diamond surface was shaded and protected against metallising prior to starting metallising with chromium.

The characteristic sections of the metallised diamonds and also of the diamond surface situated below the coating after removal by etching (removal by dissolution) or separation were examined in plan and photographed in optical and scanning electron microscope.

The thickness and surface roughness of the coatings, the depth of displacement of the interface into the depth of the diamond and the crater in separation the coating from the diamond surface and the relief of the diamond below the coating depend on the metallising temperature and time. Figure 73 shows the diagram of changes of these parameters in relation to temperature. At a metallising temperature of 1000 °C the height of protrusion of the coating above the initial surface of the diamond is equal to its thickness and amounts to approximately 1.3 μm. No displacement of the interface into the diamond was detected. When the temperature was increased to 1100 °C, the height of protrusion reached 3.8 μm, and the coating 'grew' by approximately 0.2 μm into the depth of diamond. The increase of temperature to 1300 °C (holding time was constant, 60 min) increased the height of the protruding part of the coating to 20 μm and the displacement to 2.5 μm. It should be noted that the total thickness of the chromium

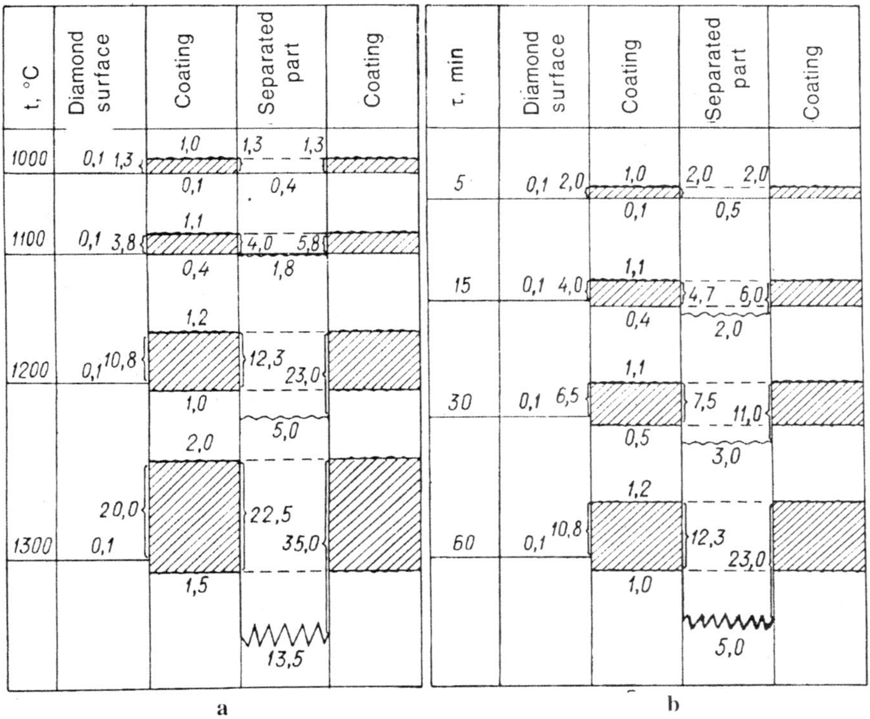

Fig.73 Diagram of the increase of the thickness of chromium coatings, displacement of their boundary with diamond and the depth of separation of chromium coatings in relation to metallising temperature at τ = 1 h (a) and time at 1200°C (b). The numbers above the surface are the values of the mean height of the relief form, the numbers around the brackets are the values of the thickness of the coating and the depth of separation of the layers (μm).

coating consists of the protrusion height and the depth of displacement into diamond and corresponds to the thickness of the coating determined from the increase of the weight of the diamond crystal after its metallising.

Increasing temperature increases the surface roughness of the coating (from 1.0 μm at 1000 °C to 2.0 μm at 1300 °C) and the roughness of the diamond below the coating and, most importantly, the depth of the separation crater greatly increases and becomes considerably larger than of the depth of displacement of the interface of the coating with the diamond. The shaded part of the diamond (zero level) does not change and its roughness is 0.08–0.10 μm.

Thus, increasing thickness of the chromium coatings with increasing temperature increases the surface roughness of the coating and of the diamond surface below the coating, the depth of displacement of the layer into diamond and the depth of the separation crater. When

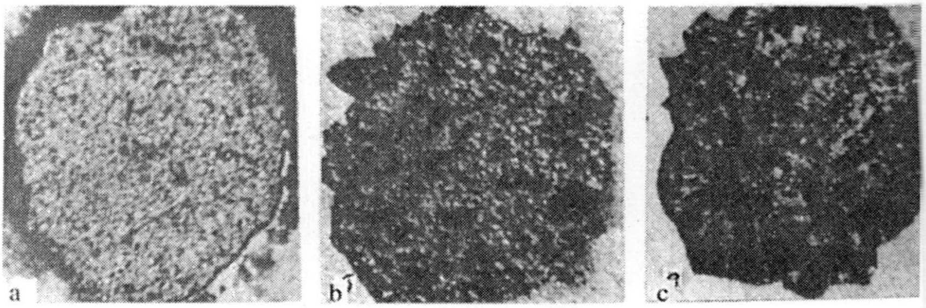

Fig.74 Photographs of spots (×30) of separated chromium coatings deposited on diamond at temperatures of: a) 1100; b) 1200; c) 1300°C.

the depth of displacement of the zone of the coating into the body of the diamond reaches 1 μm and more, the bonding strength of such a coating rapidly decreases, to almost 0.

Similar changes took place in the diamond–coating couple when increasing the contact time of diamond with the metal (Fig. 73). It is important that in these experiments the depth of the separation crater was also considerably greater than the depth of penetration of the coating into diamond. Figure 74 shows photographs produced in optical microscope of regions of the diamond surface after separation of the chromium coatings of different thickness produced at different temperatures. The photographs clearly show that increasing metallising temperature results in coarsening of the structure of separation spots and the surface of the diamond without the coating is unchanged.

A layer of chromium carbide forms at the interface between the copper–gallium–chromium alloy and diamond.

Profilographic measurements of the carbide layers of the diamond surface,[206] produced after isothermal holding the Cu–Ga–Cr melts, showed that increasing chromium concentration of the Cu–Ga melts increases their thickness and surface roughness (Fig. 75, *I*). The depth of displacement of the carbide phase into diamond also increases. For example, at 0.4% of chromium in the alloy the thickness of the carbide layer reaches 1.0 μm, and at 3.9% it is 7.0 μm; correspondingly, the depth of displacement is 0 and 2–3 μm. The holding time of the melts was constant, 15 min. Figure 75, II shows profilograms of the isothermal growth of carbide layers formed at the diamond–alloy (Cu–Ga)–0.8% Cr interface when holding the latter on the diamond surface (1000 °C) for various periods of time.

Increasing the contact time of the systems increses, as in the previous case, the thickness of the carbide interlayer, but the thickness of the interlayer is, even at τ = 60 min, smaller than when using the alloy with a higher chromium content (3.9%) but shorter holding time.

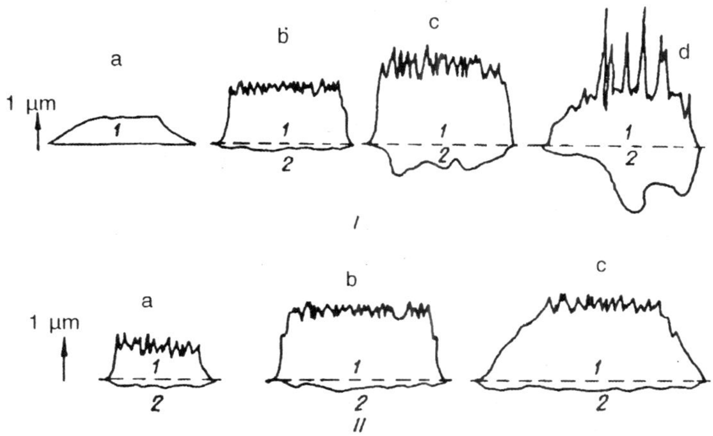

Fig.75 Profile patterns of increase (×3000) of the thickness of the carbide layer at the diamond–melt (Cu–18.8 Ga) – 0.8% Cr interface at 1000°C in relation to chromium concentration (I) (a – 0.4; b – 0.8; c – 2.0; d – 3.9%) and holding time (II) – (a – 4, b – 15, c – 60 min): 1) part of the carbide layer situated above the diamond surface; 2) part of the carbide layer displaced into diamond.

3.4.2 Effect of diffusion processes on interfacial contact strength

Analysis of these investigations makes it possible to draw the following conclusions. On the basis of general considerations it may be assumed that the strength of contact of diamond with the metal must be determined by the strength of interfacial bond of diamond (carbon) with the metal, the structure and special features of the constitution of the interfacial region. These main factors are a complex function of the chemical affinity of carbon for the metal and of the conditions of contact formation (methods, temperature, time, etc.). The experimentally determined contact strength is influenced by stresses of various origin in the interfacial region.

An important factor is the graphitisation of diamond which takes place during contact formation. The interlayer at the interface of the graphite with low mechanical strength greatly reduces the contact strength.

Attention will now be given to the dependence of contact strength on the temperature–time conditions of contact formation. The dependences of the bonding strength of diamond with metals and alloys on temperature and contact formation time, and also on the concentration of active additions has the form of curves with distinctive maxima (see Fig. 37, 38, 42, 45, 47), i.e. strength initial increases with increasing temperature and holding time and, subsequently, reaching its maximum value, rapidly decreases. The increase of strength with increasing temperature and contact time should be associated with the start and course of the process of interaction between the diamond and the metal and with the establishment of chemical bonds at the interface. These processes are accompanied by parallel processes of softening, associated with the special features of the diamond–metal interfacial reaction, which prevail at the intensification of the temperature–time conditions of contact formation. Consequently, the initially high strength rapidly decreases.

What is the nature and mechanism of softening?

It should be stressed that the fracture of a diamond–metal contact, produced at temperatures higher than optimum temperatures (ensuring the maximum strength of content) takes place through the body of the diamond crystal and not at the interface.

The decrease of strength with increasing temperature and time, displacement of the diamond–metal interface and of the fracture surface (zone) to the depth of the diamond, and the increase of the thickness of the carbide interlayer make it possible to assume that the main factor of softening are diffusion processes and the following interpretation of the softening phenomenon can be proposed. Because of the differences in the diffusion coefficients of the carbon atoms into metal D_1 and of the metal atoms into diamond D_2 at the ratio $D_1 >> D_2$ in the subsurface regions of the diamond crystal there are regions with increased vacancy concentration which can coalesce into micro- and macropores (Kirkendahl–Frenkel effect). The vacancies, micropores and dislocation disruptions in the diamond crystal, formed as a result of differences in the diffusion flows C → Me and Me → C, 'loosen up' and soften the diamond crystal.

Special attention will be given to the relationships governing the formation and softening of the diamond–metal (alloy) contact system. The coatings of refractory carbide-forming metals (Ti, V, Nb, Cr, Mo, W), deposited at relatively low temperatures (below 750–850 °C) are very thin, semi-transparent and consist (according to x-ray analysis) of mainly the metallic phase. The carbide phase is present in the form of traces. Such coatings have very low bonding strength with the diamond. Evidently, this is associated with the special features of their formation and of establishment of chemical bonds of the metal with the diamond surface.

The coatings start to form with the appearance of islands (Fig. 76,). The bond between the metal and the diamond crystal of forms only on the area occupied by the islands. Increasing temperature increases the area of the diamond occupied by the condensate, and the development of the Me + C → MeC reaction leads to the increase of the number of atoms chemically bonded with the diamond surface. This

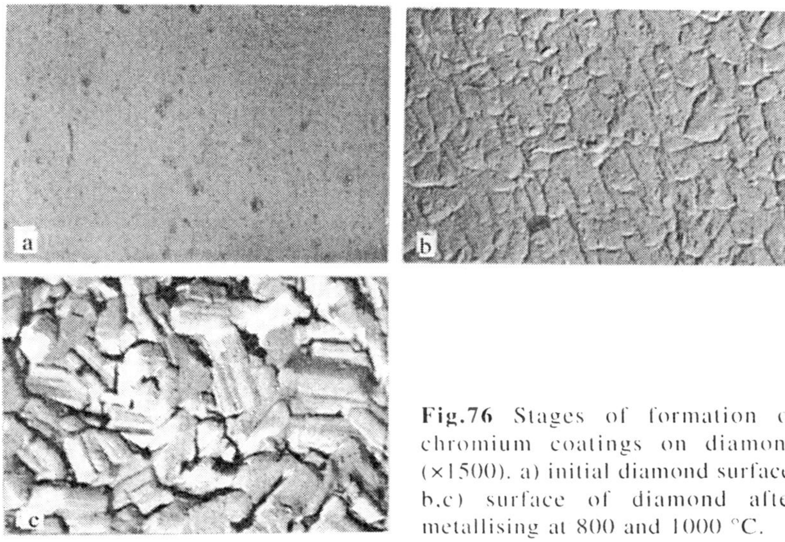

Fig.76 Stages of formation of chromium coatings on diamond (×1500). a) initial diamond surface; b,c) surface of diamond after metallising at 800 and 1000 °C.

increases the integral strength of the diamond–metal contact.

It is important to stress that separation of the coatings, deposited in this stage (rising parts of the $\sigma = f(T)$ curves) takes place mainly at the interface. In some cases, because of the high bonding strength of the diamond and the metal, fracture can take place in a small contact area in the body of the diamond. The fracture surface is light and bright in this case.

When the metallising temperature is increased further (900–1000 °C) the thickness of the carbide layers rapidly increases. Because of the active diffusion of carbon atoms into the coating (as mentioned previously), a region with higher concentration of vacancies or pores forms in the subsurface regions of the diamond. In this case, separation takes place through the weakest member of the contacting couple – through the zone with the pores, and the bonding strength is low.

In accordance with the Kirkendahl–Frenkel effect, intensification of diffusion (this is achieved by increasing temperature and holding time) results in the displacement of the contact boundary and also the growth and displacement of the zone with the pores toward the substance with the higher diffusion rate (diamond). Examination of the Kirkendahl-Frenkel effect shows that the rate of displacement of the porosity zone is usually higher than the speed of the interface.[77] This is also observed for the diamond–metal system. The fracture surface of the diamond–coating contacting couple moves into the bulk of diamond with increasing temperature and metallising time and moves away from the moving interface (Fig. 73).

The displacement of the diamond–metal interface with time is parabolic (displacement in relation to the initial level is directly proportional to the square root of the metallising time); this is typical of the Kirkendahl phenomenon and is observed in the system under examination (Fig. 77).

According to Ref. 77 and 208, the displacement of the zone of the pores is also governed by the parabolic law, although the speed of movement of this zone is higher than that of the interface. The data on the displacement of the zone with the pores (Fig. 77) and the interface confirm the validity of the concept regarding the Kirkendahl–Frenkel effect in the diamond–metal systems.

The formation of diffusion porosity should be regarded as the main reason for the decrease of the bonding strength of the metal with the diamond surface. Increasing thickness of the coating and the carbide layer increases the amount of carbon transferred into the metallic phase and results in extensive 'diffusion disintegration' of the diamond–metal contact. The 'diffusion disintegration' – pore formation – can be inhibited by arresting the diffusion flow of the carbon atoms into the metal.

According to the results of x-ray diffraction analysis, in high-temperature metallising of diamond by the contact-reaction method in the powder metallising agent, diffusion processes result in carbon saturation of not only the continuously growing coatings but also of the adjacent areas of the powder – metallising agent.

If the diamond is insulated from the metallising agent in the early stage of the diffusion in which the coatings are relatively thin and the degree of diffusion softening is negligible, further heating of such a diamond should not cause any extensive pore formation and, consequently, any decrease of the strength of contact as a result of decreasing intensity of the outflow of carbon atoms. The following investiga-

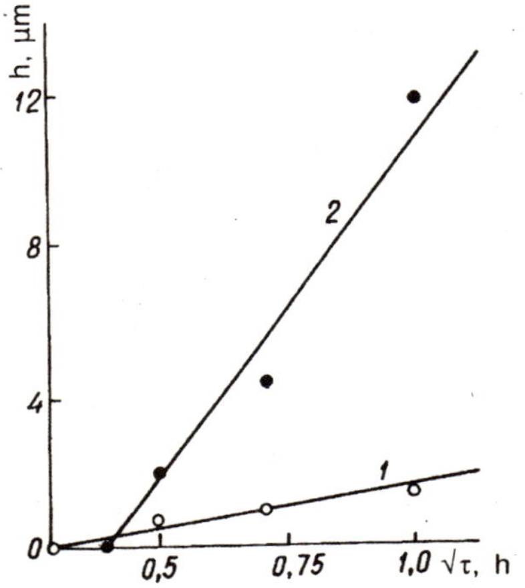

Fig.77 Depedence of the displacement into the depth of diamond of the interface (1) and the zone with pores (2) on duration of metallising with chromium at 1200°C.

tions were carried out. The diamond crystals with a thin chromium coating (h = 1.3 μm, 1000 °C, τ = 1 hour) were annealed without any metallising agent at high temperatures for various periods of time. This was followed by determining the strength of bonding of these coatings with the diamond. The experimental data basically confirm the assumption (Fig. 61 and 62) according to which the bonding strength remains high and only annealing at ≥1200 °C reduces this value from 165 to 4–10 MPa, and the coating separates along the interface. Under these high-temperature annealing conditions a significant effect is exerted by the process of graphitising the diamond surface which decreases the strength of the bond. Thus, it is necessary to accept the existence of two mechanisms of weakening of the contact between diamond and metal: diffusion disintegration of the subsurface regions of the diamond crystal, taking place at moderate temperatures (1050–1150 °C); graphitisation of the diamond surface that is in direct contact with the metal; this process takes place at high temperatures (≥1200 °C). The following experiments are interesting from this viewpoint.

The Cr coating (h = 6.3 μm) was deposited on diamond at 1150 °C for 1 hour. These deposition conditions resulted in weakening of the contact (σ = 17 MPa). It is characteristic that, under these conditions, fracture of the diamond–metal contact takes place through the body of the diamond – the zone with the pores. However, when the crystal with the coating deposited under the same conditions was annealed without the metallising agent at a higher temperature (1200–1250 °C), the contact failed through the interface, the separation surface did not penetrate into the body of the diamond and the contact spot was black (Fig. 78). The contact strength decreased to approximately 1 MPa.

Thus, the factors affecting the strength of the diamond–metal contact can be experimentally separated in certain cases.

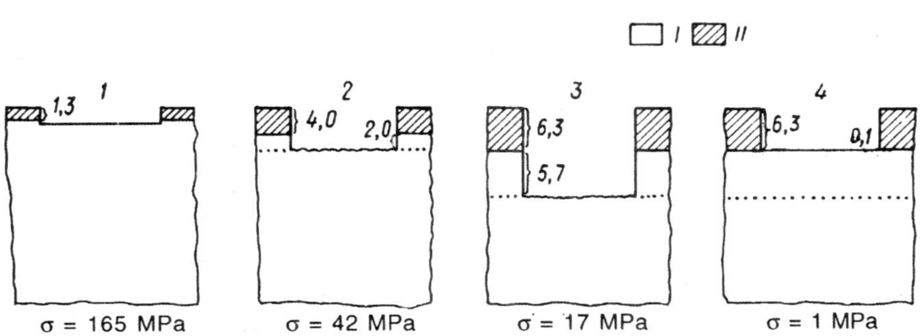

Fig.78 Diagram of separation of chromium coatings (II) deposited on diamond (I) during 1 h: 1) 1000°C (at the interface); 2) 1100°C (through the zone of its pores); 3) 1150°C (through the zone with pores); 4) 1150°C + annealing after metallising at 1200°C and 1 h (at the interface).

It should be stressed that the high-temperature deposition of Cr coatings on graphite ($T \geq 1200$ °C, $\tau = 1$ hour) in contrast to diamond does not impair the bonding strength of the coating with the substrate (see Fig. 61 and the 62). It may be assumed that the Kirkendahl–Frenkel effect operates in this system but because of the porosity of graphite and also the large length of its grain boundaries the increase of the vacancy concentration in the graphite crystals should have no effect on the mechanical strength of graphite and the strength of contact with the metal.

The processes of graphitisation of diamond and the effect on the strength of the diamond–metal contact will now be examined.

3.4.3 Effect of graphitisation on the strength of contact between diamond and metal

The diamond, being a metastable modification of carbon, can transform to graphite (graphitise) under certain conditions. Graphitisation starts on the diamond surface,[209–212] and this decreases the strength of contact of the metal or alloy with the diamond surface. This is caused by the fact that the mechanical strength of graphite (21–28 MPa)[202] is considerably lower than that of diamond and the strength of the joint cannot be higher than that of graphite. In addition, the difference in the specific volume of graphite V_{gr} and diamond V_d results in the formation of stresses in the surface layer of the diamond and, consequently, additional weakening of the contact:

$$V_{gr} = \frac{1}{\gamma_{gr}} = \frac{1}{2.25 \cdot 10^3 \text{ kg/m}^3} = 0.44 \cdot 10^{-3} \text{ m}^3/\text{kg};$$

$$V_d = \frac{1}{\gamma_d} = \frac{1}{3.51 \cdot 10^3 \text{ kg/m}^3} = 0.28 \cdot 10^{-3} \text{ m}^3/\text{kg},$$

where γ_{gr} is the density of graphite; γ_d is the density of diamond, kg/m^3.

The phase transformation of diamond into graphite during heating in the manufacture and service of diamond tools has been the subject of many investigations.[213–247] Graphitisation of natural and synthetic diamond has been examined in heat treatment under the conditions of different environments, contacting metals, etc.

It should be noted that the views on the nature of the graphite layer, deposited on the diamond surface, differ. The majority of investigators assume that diamond transforms to α-graphite, but there are data on the formation of mixtures of α- and β-forms, and in heat treatment of natural diamond even formations of coke- and soot-like substances were detected.

The authors of Ref. 213 assume that graphitisation of diamond takes place in three stages: the formation of amorphous carbon; transformation of amorphous carbon to turbostratic (hexagonal crystal system – deformed and distorted); partial graphitisation of turbostratic graphite.

The graphitisation of diamond as a process depending on many factors will be examined in greater detail.

Temperature in heating and environment

It is assumed that diamond starts to transform into graphite at temperatures higher than 1250 °C. However, the temperature of the start of graphitisation depends on the type and purity of diamond, medium, holding time.[17] According to Ref. 214–216, the surface of defect-free natural diamond with no impurities starts to graphitise in deep vacuum ($5\cdot10^{-4}$ Pa) at 1500 °C.

Synthetic diamonds with a more defective structure and impurities of relatively large amounts of metals–catalysts (1–2%) can graphitise at lower temperatures. In Ref. 217–219, graphitisation of synthetic diamond of grade AC2, AC4 and AC6 was observed at 800–1200 °C in vacuum. As the amount of impurities in diamond increased and their quality decreased, the temperature at which the phase transformation started decreased.

Nickel, iron, manganese and other metals, being catalysts of diamond synthesis and present in the diamond in the form of impurities, can subsequently become catalysts of the reversed process of transformation of diamond into graphite. When using these metals in the form of coatings, brazing alloys or binders, contacting with the diamond, graphitisation can take place, under corresponding conditions, at a considerably higher rate and at lower temperatures.

Graphitisation can also take place inside the diamond crystal.[215] In the absence of impurities and inclusions, the activation energy of the process is higher because it is necessary not only to rearrange the diamond lattice but also form a new surface. Therefore, graphitisation is observed at higher temperatures.

Investigations of the graphitisation of powders of natural and synthetic diamond in heating in an induction furnace and a resistance furnace to 1800 °C for 30 minutes (in argon) showed that the degree of graphitisation in the induction furnace is considerably higher than in the resistance furnace. This phenomenon is explained by the non-uniformly heat field formed by the high-frequency electromagnetic field.

The temperature of the start of internal graphitisation greatly decreases in the presence of inclusions.[214,215] If the surface graphitisation of diamond softens its contact with the metal, internal graphitisation may result in the failure of the crystal.

A reduction of vacuum (a decrease of the degree of rarefaction) also decreases the temperature of the start of graphitisation of diamond.[220] In air, graphite can be detected on the diamond surface already at 500–900 °C. This is the consequence of chemical interaction of diamond with oxygen of the gas medium (and not of rearrangement of the crystal lattice).[221–224]

According to a large number of investigations, the process of high-temperature oxidation of diamond takes place through the stage of its graphitisation. Since the graphitisation of diamond can take place even in the presence of small amounts of oxygen, in manufacture of diamond tools it is often recommended to use non-oxidising shielding media or high vacuum.

The literature data on the effect of the gas medium (CO/CO_2, H_2, water steam, H_2O, inert gases) are contradicting because it is very difficult to draw any unambiguous conclusions on the efficiency of application of a specific medium in preventing the phase transformation of diamond to graphite during heating. Evidently, this is explained by insufficient cleaning of the gas.

In Ref. 220, X-ray diffraction analysis and mass changes were used to determine the degree of graphitisation of natural diamond with a size of 0.05–0.07 carat. The degree of graphitisation of natural diamond in different media – argon, hydrogen, CO/CO_2 mixture and a vacuum of 10^{-2} Pa – was 0.19; 0.23; 0.29 and 4.3%, respectively.

On the basis of the experimental results, the authors have concluded that it is efficient to use argon and hydrogen as shielding media. A similar conclusion was made in Ref. 225. In heat treatment of natural diamond with a size of 0.017–0.025 carat in different media it was observed that the rate of graphitisation increases in the following sequence: vacuum – argon – nitrogen – hydrogen.[226] The degree of graphitisation was determined from the change of mass and visually. It was interesting to compare the values of the degree of graphitisation of natural diamond in argon determined by different authors using the same methods: 0.19% at 1400 °C, Ref. 220, and 0.3% at 1200 °C, Ref. 226. It can be seen that the results are almost completely opposite.

The results of examination of the graphitisation of synthetic diamond of AC2 and AC4 grade on the basis of infrared absorption spectra are presented in Table 16.

According to the authors of Ref. 217, heat treatment of synthetic diamond in argon leads to maximum graphitisation, whereas in hydrogen to minimum graphitisation. Attention should be given to the large difference in the degree, the graphitisation of diamond at 1400 °C (τ = 1 h) in hydrogen, obtained by different authors: 6% in Ref. 217 and

Table 16 Degree of graphitisation (%) of synthetic diamonds AC2 and AC4 in different media (τ = 1 h)[217]

Temperature, °C	Vacuum	Argon	Hydrogen	Temperature, °C	Vacuum	Argon	Hydrogen
	AC2				AC4		
850	3	3	3	800	3	3	3
1000	3	3	3	1000	3	3	3
1200	3	3	3	1200	3	3	3
1400	11	13	6	1400	14	14	5
1500	20	28	15	1500	19	33	16
1600	29	–	–	1600	25	–	–

0.23% in Ref. 220.

In examining the process of heat treatment of synthetic diamond of grade AC4 with a grain size of 63/50 and 200/160 less extensive graphitisation was observed in a shielding medium produced in burning activated coal (CO/CO_2).[220] The decrease of the mass in this case was 2% (1200 °C, τ = 30 min). When using dried hydrogen (dew point –50 °C) the mass loss was 3% and in moist hydrogen (dew point –10 °C) it was 16%. These values show the importance of inspecting the purity of the shielding medium in which the diamond is heat treated, in order to obtain reliable results.

In Ref. 221 it was attempted to explain the mechanism of the effect of the gas medium on the process of graphitisation of diamond, and thermodynamic calculation of equilibrium in the diamond–graphite-gas medium system was carried out. The shielding medium was oxygen, hydrogen, H_2O, water vapours and CO/CO_2. On the basis of these calculations, it was concluded that increasing heat treatment temperature of the diamond increases the probability of oxidation and, consequently, of graphitisation of diamond crystals in CO_2, water vapour and graphite. In addition, according to data in Ref. 220, at high temperatures hydrogen erodes the diamond surface thus forming the so-called etch pits with a black deposit which is evidently a deposit of graphite.

Consequently, it can be concluded that vacuum is the most suitable medium for heat treatment of diamond. Increasing vacuum decreases the probability of deposition of graphite layers on the diamond surface. On the other hand, a large increase of the pressure of the gas medium[227–229] to 2000 MPa or higher creates conditions similar to the thermodynamic stability of the diamond. The rate and degree of graphitisation rapidly decrease as well.

As already mentioned, the metallic inclusions greatly facilitate the process of graphitisation of the diamond. Taking into account the fact that, in the majority of cases, the diamond is in contact with metallic coatings and binders, it is important to know the type of changes taking place on the diamond surface (or at the diamond-metal interface in this case) in high-temperature contact of the diamond with metals and alloys.

Metals and alloys in contact with diamond

In many investigations[230–235] the authors reported a catalytic effect of the metals of the iron group on the diamond–graphite transition. In cases in which the contact of diamond with, for example, iron was 'moving', i.e. the material containing iron was constantly renewed, deposition of graphite was not detected. These conditions were created in high-temperature friction of diamond on iron (steel) during its treatment.[231–233] 'Static' contact of iron with diamond (annealing of diamond in an iron powder),[234] annealing of iron or iron-containing coatings[209] on diamond, brazing of diamonds by iron-containing brazing alloys,[81,209,235] and annealing of the diamond powder on the surface of an iron specimen[236,237]) resulted in the graphitisation of diamond crystals.

The authors of Ref. 238 carried out vacuum metallising of AC2 diamond powders with a grain size of 125/100. The coating material was copper, condensed on the diamond through a manganese sublayer. In heating diamond to 1000°C the latter was graphitised.

The interaction of natural diamond with a grain size of 5/3 with cobalt in dry hydrogen was observed at temperatures higher than 1100 °C, and with nickel under the same conditions at 800 °C.[154]

The results of electron diffraction examination of the face (111) of natural diamond–nickel coating interface, annealed at different temperatures for 5 min in a vacuum of 10^{-4} Pa, show that at 400 °C the electron diffraction patterns contain a weak, eroded graphite line.[239] The non-transition metals – tin, copper, gold, aluminium, silicon and others – do not cause graphitisation of diamond. The graphitising effect of chromium, molybdenum, and tungsten was also low.[70,153,234,235,240]

The rate of graphitisation of the diamond crystal differs for different faces. In the (110) direction, the diamond octahedron graphitises more rapidly than in the (111) and (001) directions.[215] The experimental results show[241] that there is a relationship between the degree of graphitisation

Table 17 Thickness of the graphitised layer (μm) in different gas media (T = 1500°C, τ = 60 min)

Diamond	H_2 (T_{dw} = – 20°C)	$CO + N_2$	Ar
Ground	5.36 ± 0.12	3.58 ± 0.13	2.52 ± 0.07
Polished	2.44 ± 0.07	2.28+ 0.06	0.91± 0.03

and the surface condition of the diamond grains, Table 17. Polishing of oval diamonds resulted in a decrease of the losses of diamond during heating due of graphitisation by approximately a factor of 1.9–2.5. Graphite forms on the surface of the diamond initially in the form of traces whose area then increases together with the increase of their number until the entire surface is graphitised.[215] In the initial stage, the rate of graphitisation is the highest and then decreases. The kinetics of the graphitisation process were examined in Ref. 242, 243.

In previous investigations of the temperature dependence of the contact strength of tantalum coatings with the diamond surface it was shown that even high-temperature metallising (T = 1400°C) did not result in the graphitisation of diamond and the decrease of σ was determined by high thermal stresses in the contact system.[140]

The metals can influence the process of graphitisation of diamond and, during this process, they themselves initially transform to oxides. The investigations show[225,226] that the catalytic properties in the presence of oxygen are exhibited by iron, nickel, cobalt, manganese, titanium, silver, molybdenum and boron. Tantalum is the only of the examined metals that shows no catalytic properties.

The mechanism of formation of graphite of the surface of diamond during direct contact with metals and alloys in vacuum or an inert, non-oxidising medium has not been explained. It is believed that the metal may cause direct rearrangement of the diamond lattice to graphite. Another view is associated with the assumption according to which the graphite forms through the carbide of a metal of the solution of carbon in the metal formed during interaction of diamond with the metal. For example, in Ref. 242 and 244 it was reported that the graphitisation processes can be divided into several stages: the dissolution of carbon in metals or compounds of metals followed by its precipitation in the form of graphite; the formation and subsequent dissociation of the carbide; formation of a substitutional solid solution of elements in carbon.

Graphitisation is an auto-catalytic process that accelerates with increasing number of graphitisation centres or the thickness of the graphitised layer. Diffusion dissolution of carbon reduces the thickness of the layer and, at the same time, has the reversed effect on graphitisation (reduces the rate of the process). The existence of these two limiting cases is reported in Ref. 236 and 237. From these positions, it is possible to explain the absence of graphitisation of diamond during its 'dynamic' contact with iron and the presence of an allotropic transformation of diamond into graphite in the case of 'static' interaction.

In high-temperature contact of thin (0.45 μm) and thicker (0.9 μm) niobium coatings with diamond, the first case was characterised by the graphitisation of diamond and, consequently, by a large decrease of the strength of the diamond–metal contact. In the second case, there was no graphitisation and the strength was relatively high.[139] The carbidisation of thicker coatings requires a larger amount of carbon and, consequently, there was no deposition of graphite layers in the second case, and the resultant graphite was removed by diffusion. Both 'dynamic'and 'static' contacts were observed in these cases.

The authors of Ref. 119 deposited a molybdenum coating on diamond in heating to 1000 °C. When the coating consisted of the intermediate Mo_2C carbide and could bond carbon, no graphite interlayer ('dynamic' contact) was detected at the interface with diamond. However, when Mo_2C transformed with time to the higher carbide MoC and the diffusion rate decreased, the graphite layer was found at the interface ('static' contact).

The metals which efficiently dissolve carbon (cobalt, nickel, manganese, etc) can act, under certain conditions, as catalysts of the process of graphitisation of both diamond and other metastable forms of carbon (for example, carbon films, fibres).[245–247] To prevent catalytic graphitisation, it is necessary to block the process of interaction of diamond with the metal or arrest the precipitation of the carbon atoms from the solution or the carbide (if it is assumed that the graphite forms through the intermediate carbide phase or a solid solution).

The metals inert with respect to diamond or metals – solvents of carbon, saturated in advance with carbon, should not cause diamond graphitisation. There are no reliable experimental data in this respect. Some authors indicate the following possibility: the saturation of the iron coating on the diamond with carbon[248] evidently decrease the degree of of graphitisation of diamond. The decrease of the solubility of carbon into the nickel by adding silicon (to 40%) prevented the graphitisation of diamond at a temperature of 1000 °C.[70] Evidently, the strong carbide-forming elements also do not cause the graphitisation of diamond. It is important to carry out special examination of the catalytic effect of metals on the graphitisation of diamonds and the decrease of contact strength.

In Ref. 209, an interesting phenomenon was observed when investigating the strength of diamond–metal contact. When a rod was made of tungsten or molybdenum, the bonding strength of the coatings deposited under the optimal conditions with diamond was high. For example, the bonding strength in the case of the diamond–molybdenum coating–molybdenum rod brazed joint was 130 MPa. If the molybdenum rod was replaced, under the same conditions, with an iron (steel) rod, the bonding strength decreased to the values close to 0, and the contact spot of the diamond surface had strong black colour.

A similar effect was also detected in the case of the diamond-adhesion-active alloy Cu–Ga–Ti, Cu–Ga–Cr contact couple, and also when adding a small amount of iron or nickel into the brazing alloy. This phenomenon was very important and subject to special investigations. The data are presented in Table 18 and 19 and on tensile test diagrams, Fig. 79, of molybdenum and iron rods brazed with the Cu–18.8% Ga brazing alloy to the molybdenum–metallised diamond surface.

Comparison of the experimental results on the effect of iron and nickel on the strength of the diamond–metal contact with the data on the effect of these metals on the graphitisation of diamonds indicates that in the process of formation of the bond the surface of the diamond was graphitised and this resulted in a large decrease of the bonding strength of diamond with the metal.

The quantitative effect of iron and nickel as catalysts of graphitisation of the process of softening of the diamond–metal joint has been investigated using the following procedure.[209] Charges of previously prepared alloys (Cu–Ga) with additions of 0.05; 0.1; 1.0; 5.0% Fe and also 0.05; 1.0; 5.0% Ni, were deposited on the molybdenum-metallised surface of diamond. The specimens were then held in a vacuum furnace at 1000 °C for 0.5 hours and, subsequently, a molybdenum rod was brazed to the spreading droplet at temperature slightly lower than the previous temperature (990 °C).

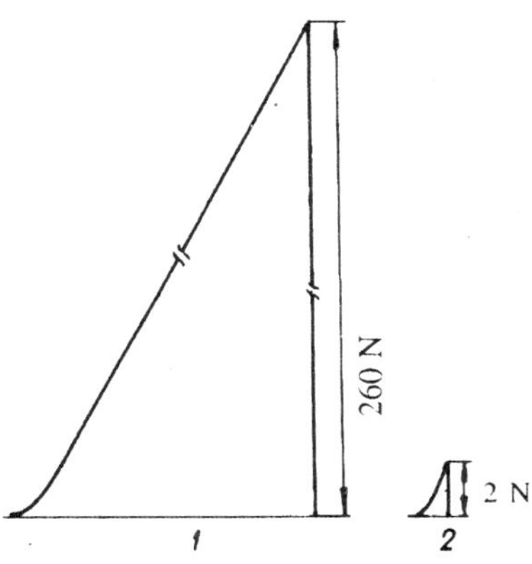

Fig.79 Diagrams of tensile strength of molybdenum (1) and iron (2) rods, brazed with Cu–18.8% Ga brazing alloy to molybdenum-coated surface of diamond (σ_1= 130 MPa, σ_2 = 1 MPa).

Table 18 Strength of bonding of the Mo coating with diamond in relation to the material of the rod and the composition of the brazing alloy

Material of the rod	Composition of the brazing alloy, %	Brazing temperature, °C	σ, MPa	Surface condition of diamond after separation of the coating
Mo	Cu	1120	129	Bright spot
	Cu – 72.0 Ag	830; 1000	151	" "
	Cu – 18.8 Ga	1000; 1100	129	" "
	Cu – 10.0 Ni	1120	0.1	Black spot
	Cu – 5.0 Fe	1120	0.1	" "
	(Cu – Ga) – 5.0 Fe	1000	0.3	" "
Fe	Cu	1120	0.1	" "
	Cu – 18.8 Ga	1000; 1100	0.2	" "
	Cu – 72.0 Ag	1000	0.3	" "
	Cu – 72.0 Ag	950	0.5	" "
	Cu – 72.0 Ag	850	151	Bright spot

The results of tests of the strength of the diamond–coating Cu–Ga–Fe (Cu–Ga–Ni) alloy–molybdenum rod contact system are presented in Fig. 80. With increasing amount of iron in the brazing alloy the bonding strength of the coating with the diamond decreases from 129 to 1 MPa and lower. It is characteristic that even a small addition of iron in an amount of 0.1% decreases the maximum bonding strength to 20 MPa. In addition, as the iron content of the alloy increases and the bonding strength correspondingly decreases, the contact spot on the diamond surface becomes darker.

Nickel also reduces the bonding strength of the coating with diamond (see Fig. 80). However, its effect is weaker than that of iron. For example, an addition of 5.0% nickel to the brazing alloy decreases the contact strength only to values of the order of 30–40 MPa.

The conditions of the experiments described were then slightly changed. Iron coatings of different thickness were deposited on the molybdenum-metallised surface of diamond by electron beam spraying. The coatings were then annealed in a vacuum at 1000 °C for 0.5 hours (the annealing conditions were identical with those used previously). The molybdenum rod was brazed with a Cu-Ga alloy. The de-

Table 19 Strength of bonding of several alloys with diamond in relation to their composition and rod material (brazing temperature 1000°C)

Material of the rod	Composition of the alloy, %	σ, MPa	Condition of the surface of diamond after separation
Mo	(Cu – Ga) – 0.5 Cr	108	Bright spot
	(Cu – Ga) – 1.0 Ti	312	" "
	(Cu – Ga) – 0.5 Cr – 5.0 Fe	0.4	Black spot
	(Cu – Ga) – 1.0 Ti – 5.0 Fe	0.4	" "
	(Cu – Ga) – 0.5 Cr – 10.0 Ni	0.3	" "
	(Cu – Ga) – 1.0 Ti – 10.0 Ni	0.4	" "
Fe	(Cu – Ga) – 0.5 Cr	0.2	" "
	(Cu – Ga) – 1.0 Ti	0.2	" "
	(Cu – Ga) – 0.5 Cr – 5.0 Fe	0.1	" "
	(Cu – Ga) – 1.0 Ti – 5.0 Fe	0.1	" "
	(Cu – Ga) – 0.5 Cr – 10.0 Ni	0.1	" "
	(Cu – Ga) – 1.0 Ti – 10.0 Ni	0.1	" "

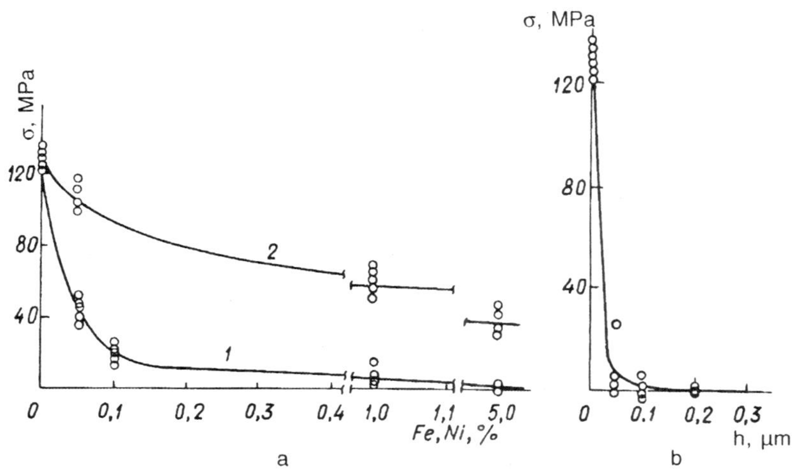

Fig.80 Dependence of the bonding strength of molybdenum coatings with diamond on the content of iron (1) and nickel (2) in a copper–gallium brazing alloy (a) and the thickness of the iron film deposited on the coatings (b).

termined bonding strength of the coating with diamond decreases with increasing thickness of the iron film, Fig. 80. The form of the curve is identical with the alloys containing iron.

To confirm the graphitisation mechanism of softening of the diamond-metal contact, investigations were carried out into the effect of the composition of the brazing alloy and the rod material of the bonding strength of the molybdenum coating (Table 20) or the adhesion-active alloy with the graphite surface. The values of the bonding strength with diamond of several adhesion-active alloys with additions of iron and nickel at 1000 °C are presented below:

Composition of alloy, %	σ, MPa
(Cu–Ga)–0.5Cr	22–25
(Cu–Ga)–0.5Cr–5.0Fe	15–20
(Cu–Ga)–0.5Cr–10.0 Ni	15–20
(Cu–Ga)–1.0 Ti – 5.0 Fe	15–20
(Cu–Ga)–1.0 Ti – 10.0 Ni	15–20

It can be seen that the bonding strength of the adhesion-active alloy

Table 20 Values of the bonding strength for molybdenum coatings with graphite in relation to the rod material and the composition of the brazing alloy

Rod material	Composition of the brazing alloy, %	Brazing temperature, °C	σ, MPa
Mo	Cu	1120	22 – 25
	Cu – Ga	1000	22 – 25
	Cu – Ga	1100	22 – 25
	(Cu – Ga) – 5.0 Fe	1000	15 – 20
	(Cu – Ga) – 10.0 Ni	1000	15 – 20
Fe	Cu	1120	15 – 20
	Cu – Ga	1000	15 – 20

Comment: Deep separation from graphite surface was recorded.

and the molybdenum coating with the graphite surface slightly decreases in the presence of iron (nickel) in the brazing alloy or when the molybdenum rod is replaced with an iron one (Table 20) but remains sufficiently high and not close to zero, as in the case of the system with diamond. Thus, the strength of the diamond–metal contact is, as a result of graphitisation of diamond in the interfacial region, considerably lower than the strength of the contact produced using the same metal or alloy and under the same conditions but with graphite.

Thus, the strength of the diamond-metal contact decreases as result of graphitisation not only (and, obviously, not as much) as a result of the appearance of a mechanically weak graphite interlayer between the diamond and the metal but also as a result of high structural stresses developed in the near-contact region of the diamond during its transition to graphite which has a considerably larger specific volume.

3.4.4 Effect of the gas medium on the strength of the diamond-metal contact

Fracture of the surface of diamond and its graphitisation are strongly influenced by the gas medium in which the diamond is heated, although this effect is weaker than that of the interaction with metals. The most intensive failure of diamond is observed when heating in air, in hydrogen and halides. Oxygen, hydrogen and halides, interacting with carbon, form volatile chemical compounds.[249] The resistance of diamond in the shielding medium of inert gases (nitrogen, argon, He) is higher.

However, it should be taken into account that even shielding gases (without mentioning air) contain oxidising components, and even a small amount of these components does not make it possible to prevent completely the possibility of erosion of the diamond surface and its graphitisation. The metallic coatings deposited on diamond can reduce the extent of failure of the diamond surface or, on the other hand, can increase the extent of failure.

The authors of Ref. 250 observed and investigated a very rapid, catastrophic decrease of the bonding strength, especially of molybdenum coatings with the diamond surface, annealed in hydrogen. The contact strength decrease to almost zero.

The metallic coatings were deposited on diamond using the method discussed previously by annealing in a powder metallising agent in a vacuum $(2–5)\cdot 10^{-3}$ Pa under different temperature–time conditions. The thickness of the coatings was 0.5–2.0 μm.

Metallised diamond crystals (1–1.5 carat) were annealed in hydrogen for industrial cleaning (dew point –20 °C) and in dry hydrogen with a dew point of –50 °C. The coatings of molybdenum, chromium and tungsten were used in the pure form and combined coatings of these metals and copper were also investigated.

The results of measurements of the bonding strength of the coatings with the diamond surface are presented in Table 21; the effect of the annealing time of the coatings on diamond in hydrogen at dif-

Table 21 Bonding strength of coatings with diamond annealed in hydrogen

Coating	σ, MPa	
	Initial	After annealing in hydrogen (t = 1100 °C, τ = 30 min)
Cr	165	155
Mo	130	0
Mo (inner layer), Cu (outer layer)	130	64
Cr – Mo – Cu (polymetallic)	69	56

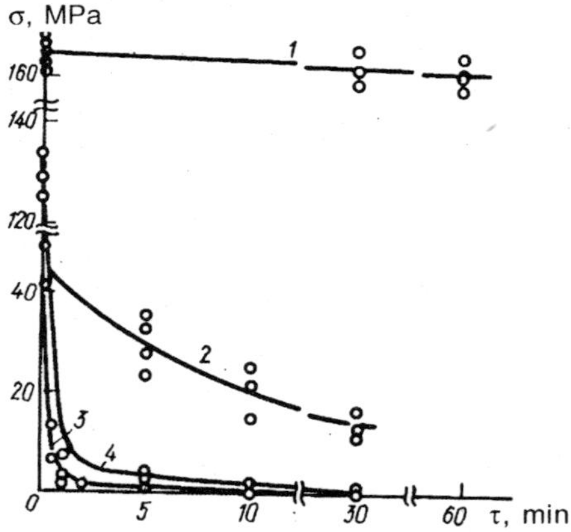

Fig.81 Dependence of the strength of bonding of molybdenum, tungsten and chromium coatings with diamond on annealing time in H_2 gas media at atmospheric pressure. 1) Cr, 2) W, 3) Mo at 900°C, H_2 dew point –50°C, 4) Mo at 800°C, H_2 dew point –20°C.

ferent temperatures on the strength characteristic of the coatings is presented in Fig 81.

The results show that in the case of the molybdenum coating, deposited in vacuum at 950°C for 60 min, with a bonding strength to diamond of approximately 130 MPa, annealing in hydrogen (rapidly dried) at 900°C reduces this parameter to zero. At the same time, the bonding strength of the tungsten coating under the same conditions was only halved (from approximately 40 to 20 MPa), whereas the chromium coating efficiently protected diamond and its bonding strength remained almost unchanged.

It can be assumed that the softening of the diamond–metal contact is caused by the hydrogenizing of the carbon of diamond in the hydrogen medium that is intensified by the presence of the metal acting as a catalyst.

The literature data[251–253] show that annealing of diamond in hydrogen is accompanied by the formation of hydrocarbons. According to the data in Ref. 252, if the diamond is in contact with nickel foil, the hydrogenizing rate increases seven times, iron foil increases the hydrogenizing rate 3.5 times, and platinum foil 1.5 times.

We have investigated the hydrogenizing of diamond (determined the mass loss of a diamond powder charge) when the metal was on its surface in the form of a coating (thickness 0.5–1.5 μm), deposited under the same conditions as in the case of diamond single crystals used for determining the bonding strength.

Figures 82 and 83 show the results of examination of the mass loss of diamond powders of 315/250 and 80/63 μm particles in annealing in dried hydrogen. The results show that the mass loss of the diamond with the coatings is higher than that of the non-metallised diamond (we assume that this is due to the acceleration of hydrogenizing), and that the loss is maximum for the molybdenum coating.

The mechanism of the process of weakening of the diamond–metal contact requires detailed examination. The following mechanism can be proposed. Since the diffusion coefficient of hydrogen in the metal is higher than that of carbon (for example, for molybdenum at 900 °C, the coefficient of hydrogen diffusion is 10^{-5}–10^{-6} and that of carbon is 10^{-7}–10^{-8} cm^2),[254,255] the interaction of hydrogen with carbon for the

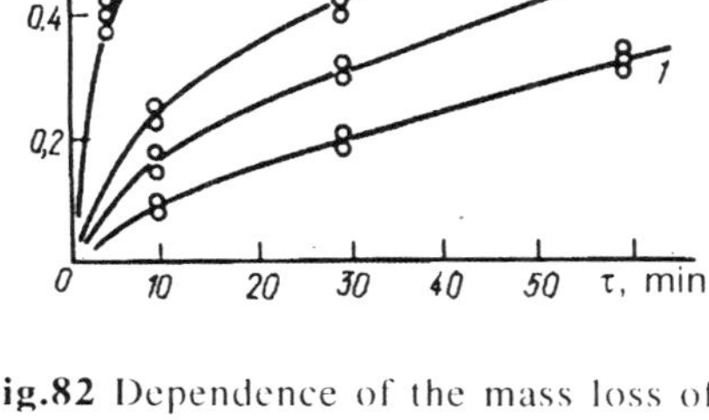

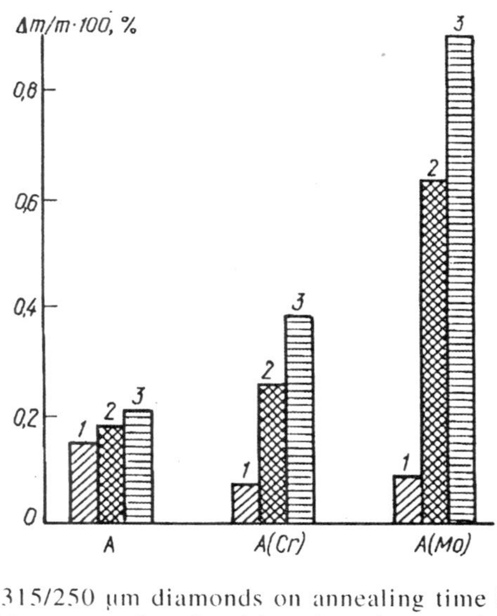

Fig.82 Dependence of the mass loss of 315/250 μm diamonds on annealing time in strongly dried hydrogen (1000 °C). 1) without a coating; 2) with a chromium coating; 3) with a tungsten coating; 4) with a molybdenum coating.

Fig.83 Diagram of the mass loss of diamonds annealed at 900° (τ = 30 min) in vacuum of $2{\cdot}10^{-3}$ Pa and in H_2 without a coating and metallised with chromium and molybdenum: 1,3) 80/63 μm at $2{\cdot}10^{-3}$ Pa in vacuum (1) and in hydrogen (3); 2) 315/250 μm in hydrogen.

diamond–metal film–hydrogen system will take place preferably at the diamond–metal interface. Hydrogen fractures the bonds of metal with carbon, for example, by the mechanism Me–C+H→C–H+Me with the formation of methane. At a hydrogen surplus and at comparable values of the energy of the Me–C and C–H bonds, the reaction takes place from left to right and the Me–C bonds fracture, thus disrupting the bonding of the film of the metal with the diamond. This was also detected for the molybdenum and, to a lesser extent, tungsten coatings.

The adhesion of the chromium film to diamond should be less subjected to the softening effect of hydrogen because of the stronger chromium carbide and the higher energy of the Me–C bond. Evidently, this position can be generalised: the metals with high chemical affinity for C should be less subjected to hydrogen softening. The role of metal in hydrogenizing of the carbon of the diamond is reduced mainly to the atomisation and ionisation of hydrogen.

The softening effect of hydrogen can be reduced using polylayer coatings, in which the outer layer consists of the metal with low permeability for hydrogen or in which hydrogen is not soluble in the form of ions (for example, copper or silver). For the coatings where the internal layer, facing the diamond, consists of molybdenum, and the outer layers are made of copper, the adhesion of the coating to the diamond is higher after annealing in hydrogen.

In practice, the application of chromium coatings on the surface of the diamond, preferred from the viewpoint of the small effect of hydrogen softening, requires clean gas media (strongly dried hydrogen) because of the formation of difficult-to-reduce oxides on the surface which complicate wetting and spreading of the brazing alloys and binders of the diamond composition. The use of such gas media under industrial conditions is associated with difficulties. The molybdenum coating, whose oxides are easily reduced by the hydrogen used in industrial cleaning when suppressing the effect of the decrease of adhesion of the metallic film to the diamond, can be widely used for industrial technologies.

3.4.5. Effect of thermal stresses on the strength of the diamond–metal contact

The internal stresses play an important role in the group of various mechanical factors having a negative effect on the strength of the diamond–metal contact. Investigators subdivide these stresses into groups:[256–259]

1) The stresses due to the noncoherentness of crystallographic lattices of the substrates in contact (this arises in the processes of solidification of alloys or condensation of coatings from the gaseous phase);

2) The thermal stresses due to the thermal expasion mismatch of dissimilar materials.

The stresses reduce the bonding strength and cause fracture of the joints in dissimilar materials in service. In particular, at tensile stresses exceeding the permissible limits, the coatings separate from the flat or convex substrate, and cracks form in them. Therefore, with a small number of exceptions, it is necessary to take measures to reduce the stresses. The following measures should be mentioned: 1) matching the properties of the coating and the substrate; 2) regulation of the thickness of the coating; 3) formation of an intermediate layer with matched properties between the coating and the substrate; 4) control of the structure the coating.

1) the matching of the properties of the coating and the substrate reduced the difference in the thermal expansion coefficients α_c, α_s (α_c is the coefficient of thermal expansion (CTE) of the coating, α_s is the CTE of the substrate). At $\alpha_c = \alpha_s$ no stresses form in the coating and the substrate. To prevent the formation of highly detrimental tensile stresses, in practice attempts are made to retain to a certain degree the inequality $\alpha_c < \alpha_s$. However, matching of the CTEs does not always make it possible to obtain the required strength of the brazed joint. It is desirable to reduce the elasticity modulus of at least one of the joined materials E_c or E_s and the Poisson coefficients.

To determine the thermal stresses, formed in the coating during cooling of the system after high-temperature condensation, the equation was derived in Ref. 257 for the elastic deformation range

$$\Delta\sigma_c = \frac{E_s E_c \Delta\alpha \Delta T}{(1-\mu_c)E_s + \frac{h_c}{h_s} E_c (1-\mu_s)}, \tag{3.6}$$

where E_c, μ_c, h_c is the elasticity modulus, the Poisson coefficient, and coating thickness, respectively; E_s, μ_s, h_s is the elasticity modulus, the Poisson coefficient, and half thickness of the sheet (substrate), respectively; $\Delta\alpha$ is the difference of the coefficient of thermal expansion of the coating and the sheet in the temperature range ΔT.

Consequently, the stresses in the sheet will be

$$\Delta\sigma_s = -\Delta\sigma_c \frac{h_c}{h_s}. \tag{3.7}$$

In the formation of metallic coatings a significant effect is exerted by the plastic deformation capacity of the metal,

2) Regulation of coating thickness. When the stresses in the deposited coating reach critical values, exceeding the bonding energy, the

coating can separate from the substrate. The values of the energy, generated as a result of stresses, depends on the coating thickness[258]

$$W_{in} = \left(\sigma_{in}^2 / 2E\right) Sh, \tag{3.8}$$

where σ_{in} are the internal stresses in the film; h is the thickness of the film; E is the Young modulus, GPa.

Equating the energy of adhesion to the energy of stresses allows to evaluate (very approximately) the minimum threshold thickness of the coating

$$h \leq \left(2W_a / \sigma_{in}^2\right) E, \tag{3.9}$$

W_a is the energy of adhesion.

As one can see, the values of the stresses strongly affects the evalue of h; increase of σ results in a decrease of h.

3) The formation of an intermediate layer with the matched properties between the coating and the substrate is possible as a result of depositing an additional layer on the outside or as a result of atomic or mutual reaction diffusion. The role of the intermediate layer as a means of reducing the stresses is explained in Fig. 84.[259] The vertical coordinate in Fig. 84 shows the values α of the CTE of the substrate material (M), the intermediate layer (IL) and the coating (C), and the horizontal coordinate gives the direction normal to the surface: a) there is no intermediate layer and a sharp jump α is detected; b) the intermediate layer is present and α is an intermediate value $\alpha_c < \alpha_{il} < \alpha_m$.The magnitude of the jump of α is smaller. Due to the small thickness of the intermediate layer and the occurrence of diffusion processes, a favourable case can be realised, as shown in Fig. 84b: α changes smoothly. However, if the mechanical strength of the

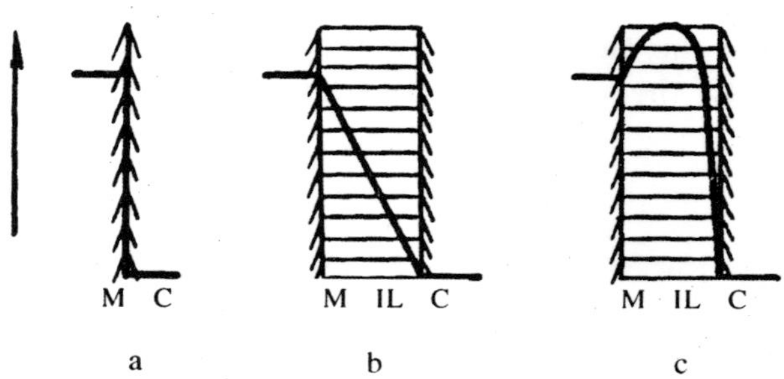

Fig.84 Effect of the intermediate layer on stresses.

layer is low, its effect becomes detrimental; c) the formation of the intermediate layer has a detrimental effect on the bonding strength since the difference $\alpha_{il}-\alpha_{c}$ is larger in this case than the initial difference.

4) Control of the structure of the coating. By influencing the structure of the coating, it is possible to control the stresses. When coatings are deposited from the gaseous phase, the stresses can be reduced by methods such as preheating the substrate, annealing deposited coatings, variation of the condensation rate, etc.

When brazing diamonds in producing tools where it is possible to avoid diffusion softening of the contact and surface graphitisation of the diamonds crystal, it is also possible to reduce the thermal stresses at the interface.

The coefficient of thermal expansion of the diamond in the temperature range from room temperature to 1000 °C ($\alpha_A=(2-5)\cdot10^{-6}$ deg^{-1}) is considerably lower than in the case of metals and metallic alloys ($\alpha_{Me} = (8-20)\cdot10^{-6}$ deg^{-1}). Therefore, the variation of the temperature of the diamond–metal contact is accompanied by the formation of stresses in this contact. These stresses are often so high that they cause fracture of the diamond crystal or greatly reduce its effective strength. (The external loads, applied to the diamond, are added to the internal stresses and the joint fractures at a low value of the external force). The shear stresses, formed when bonding the diamond and the metal on the plane, are most dangerous.

The stresses in the brazed joint can be reduced or eliminated using sufficiently ductile brazing alloys. Because of plastic deformation of the brazing alloy characterised by a low yield limit, the resultant stresses relax and, conversely, increasing hardness of the brazing alloy decreases the strength of the brazed joint.

Quantitatively, this problem has been examined by special investigations of a copper–gallium–tin brazing alloy. The alloy of copper with Ga (18.8% Ga) has sufficiently high ductility (δ = 25.1%). The addition of tin greatly embrittles this alloy because of the formation of intermetallic compounds Cu_3Sn, Cu_3Sn_8, etc.

Prior to brazing, diamond was metallised with chromium; brazing was carried out in a vacuum of 10^{-3} Pa. The strength of the brazing contact between the flat diamond surface and the end of the molybdenum rod, determined in the tensile test, was compared with the mechanical properties of the alloy with different tin content. The data are presented in Fig. 85 and Table 22 which gives the physical–mechanical properties of the alloy, the contact strength, and the calculated values of the residual stresses in the brazed joint.

The increase of hardness HRB of the alloy from 33 to 67 halves the strength of the contact. A further increase of strength has a less

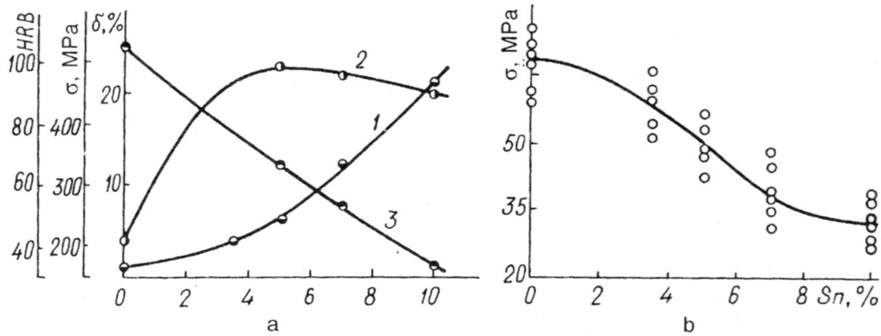

Fig.85 Effect of the tin content on the mechanical properties of alloys (Cu–18.8% Ga)–Sn (a) and strength of bonding (b) of the chromium coating to diamonds when brazing Mo rod to the coated surface with these alloys: 1) hardness; 2) tensile strength; 3) relative elongation.

Table 22 Mechanical properties of CuGaSn alloys

Sn content in Cu – 18.8 Ga alloy, %	HRB Hardness	Relative elongation δ, %	Relative reduction in area ψ, %	Poisson coefficient μ	Tensile strength σ_B, MPa,	Elasticity modulus *E*, MPa
–	33	25.1	13.2	0.53	212	845
3.5	41.5	–	–	–	–	–
5	49	12.0	2.7	0.23	492	4100
7	67	7.8	2.9	0.37	480	6154
10	94	1.0	0.4	0.4	453	45300

marked effect.

As already mentioned, embrittlement of certain brazing alloys can also be detected when they dissolve chemically active elements, for example titanium. The addition of titanium to copper greatly increases the brittleness of the alloy and, consequently, only low concentrations of this component in the brazing alloy are permissible (0.8–1.2%). The mechanical properties of several adhesion-active alloys are presented in Table 23.

Thus, the mechanical properties of brazing alloys for brazing diamond must satisfy a number of requirements. The strength characteristics of the brazing alloys must be sufficiently high to ensure fixing of the diamond under the service conditions of the component with the effect of working loading on the diamond element. The brazing alloy must also be sufficiently ductile to ensure stress relaxation in the brazing alloy–diamond interfacial region – these stresses form during tempera-

Table 23 Hardness of some adhesion-active copper-based alloys

Chromium concentration in Cu–18.8% Ga alloy, %	Titanium concentration in Cu–18.8% Ga alloy, %	HRB hardness of the alloy	
		(Cu–Ga)–Cr	(Cu–Ga)–Ti
0	0	33.0	33.0
0.2	–	41.7	–
–	0.37	–	52.7
0.8	0.74	42.7	59.5
2.0	–	49.2	–
–	2.2	–	73.0
3.9	3.6	51.0	83.8

ture changes (cooling after brazing) because of the difference of the values of the coefficients of thermal expansion of diamond and the metallic brazing alloy.

CHAPTER 4

BRAZING OF DIAMONDS

4.1 MAIN TYPES OF DIAMONDS COMPONENTS

The components containing diamonds are subdivided into the following groups:

1) Diamonds tools;

a) single crystal – with large diamond cutting elements – single crystals (from several fractions to 1–2 carat) or polycrystalline cakes (the size to several millimetres) secured in a holder;

b) abrasive – with fine diamond particles (~100 μm) or larger diamond grains (400–900 μm and larger) secured in the matrix–binder.

2) Diamond elements of electronic devices – thermistors, wheres diamond is the working body with specific electrical properties and is connected to metallic electrodes; heat sinks in which the diamond element, characterised by high thermal conductivity, is connected to the base of the electronic circuit, and some other components.

3. Diamond components used in sound reproduction technology (diamond needles of sound recorders), jewelry components, etc.

In all these cases, it is necessary to ensure strong fixing of the diamond working element in the holder or matrix of the component. The design and shape of the joints between the diamond element and the base will now be examined.

Single crystal diamond tools

Various diamond tools – cutters, glass cutters, dies, tips are produced from single crystals of diamond or their particles. In most cases, natural crystals, 0.1–2 carats, are used, and in individual cases larger diamonds can also be used.

Industry produces various types of diamond cutters. Machining with these cutters is carried out in most cases in machines with high accuracy and high cuttings rates, small feed and cutting depth. These cutters are passing, undercutting, boring and are used for finishing machining of components made from nonferrous metals.

The existence of the large number of shapes of components processed by grinding, the wide range of requirements on the machined surfaces, the large number of dimensions and characteristics of abrasive wheels to be straightened, etc. determine the wide range of the diamond straightening cutters used at present. The main types of diamond cutters are shown in Fig. 86.

The design and application of the cutters made from diamond polycrystals (ballas and carbonado) do not differ greatly from those examined above. These tools operate under severe conditions characterised by the formation of relatively high cutting forces and temperature in the working zone. The high forces act on the cutting tool also in the processes of sharpening the cutter – initial and subsequent. The design of the joint section represents the metallic component with a blind hole into which the diamond part is inserted. Prior to brazing it is necessary to secure the diamond part by caulking – compressing the metallic component.

The characteristic feature of the contact section for many of the component examined previously is the highly nonuniform and often noncapillary gap between the metallic component and the diamond component which has no distinct shape. Therefore, during brazing the component must be vertical with the open part of the fitting hole facing upwards.

It should be noted that at the small dimensions of the diamond element and the small fitting hole, the brazing gaps are small and can be regarded as capillary. In certain cases, the joint between the dia-

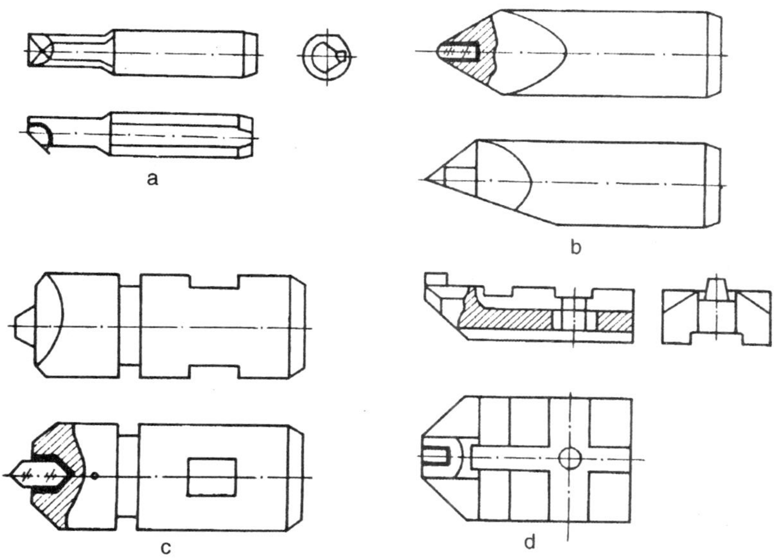

Fig.86 Types of diamonds cutters: a) lathe turning cutters for continous holes; b) dividing; c) straightening; d) for profiling worm grinding discs.

mond crystal and the base of the tool is produced on a plane, and the brazing gap can be set in this case.

The authors of Ref. 260 and 261 considered the results of calculations of mechanical stresses in diamond grains–metal matrix system when producing tools by liquid phase technology. The authors proposed models describing the interaction of a diamond grain with a meniscus of the brazing alloy and, using these models and taking into account the non-uniformly stress-strained state of the brazing alloy they explain the mechanism of retention of the diamond grain on the working surface and investigated the effect of the properties of the brazing material on diamond retention. Consequently, it was possible to select the optimum composition of the brazing alloys.

A unique method of producing diamond tools was proposed in Ref. 262. They used cast iron to fix the diamond in a cutting tool, because during rapid cooling and solidification this material becomes very hard and ductile and is a good binding material. The cast iron was poured in at 1400 °C. The experimental results show that this is accompanied by partial transformation of diamond to graphite and by its dissolution in the melt, by the formation of microcracks filled with liquid cast-iron and by bonding of the carbon of the diamond with iron (formation of iron carbide). The authors stress that these reactions ensure strong bonding of the diamond in the holder of the cutting tool, although this is doubtful taking into account the surface graphitisation of diamond.

Recently, in addition to diamond single crystals, tools have also been produced using diamond polycrystals. For example, the authors of Ref. 263 proposed polycrystalline Dipax-diamond which, according to the authors, has higher properties than Syndite-diamond and other polycrystals. The Dipax tools operate efficiently in processing materials without iron: multiphase ceramics, high abrasion-resistant alloys, plastics, fibre-reinforced polymers and other materials with low machinability.

Figure 87 shows several types of tools fitted with Dipax diamond.

The authors of Ref. 264 described the fine-grained grade of Syndite developed especially for the treatment of abrasive wood composites (Fig. 88).

The durability of tools based on this type of diamond is 30-240 times higher than that of the tools with hard-alloy insert. This is due mainly to the fact that the wear of diamond polycrystals does not take place by the erosion mechanism (for example, twisting, shear, which is typical of the hard alloy, especially WC–Co), but by the mechanism of micro-cleavage of the cutting edge, i.e. its cutting capacity is continuously restored.

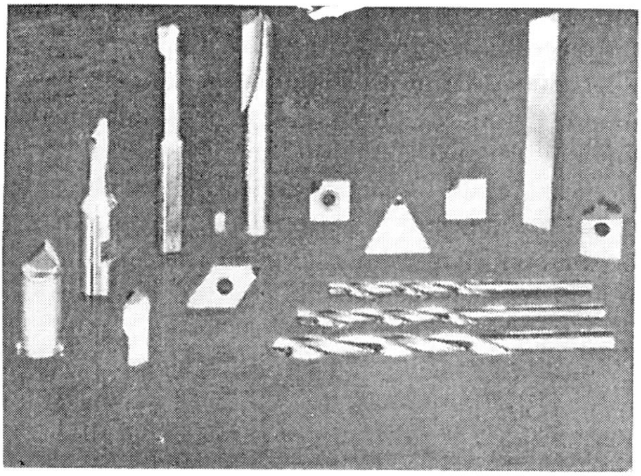

Fig.87 Types of tools with Dipax diamond.

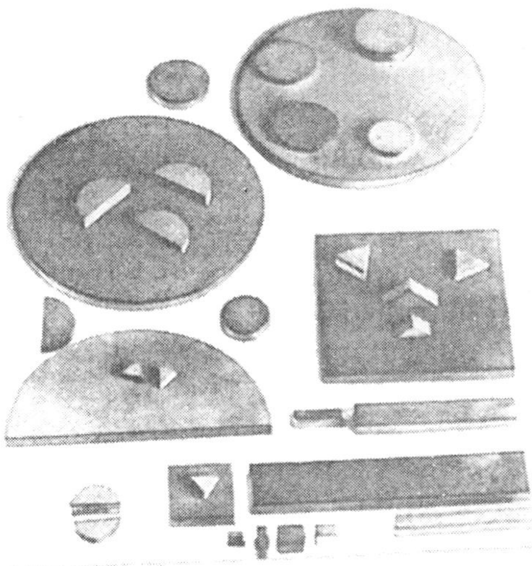

Fig.88 Types of components made of Syndite.

Abrasive diamond tools

A diamond tool with fine diamond particles or larger diamond grains has the form of grinding wheels of various shape and size (using metallic binders), drilling crowns, straightening rods, etc. These tools are produced mainly by powder metallurgy methods, especially by liquid-phase sintering or impregnation. In liquid-phase sintering, the diamond-bearing layer is pressed and then sintered at the temperature of formation of the liquid phase.

The following the procedure is used to fix the diamond element in the tool produced by impregnation. The diamond grain is pressed into the matrix produced from a refractory metal powder, and this is followed by impregnating the pressing with the liquid–metal binder–brazing alloy at temperatures lower than the melting point of the 'frame' component. The brazing alloy reaches the surface of the diamond, produces the 'socket' of a joint and interacts with it. In the drilling single-layer crowns, the diamond is not pressed completely into the matrix and part of the diamond grain protrudes abouve the working surface.

Diamond components of electronic devices

This group of diamond components includes heat sinks in which the diamond crystal in the form of a sheet or a parallelepiped (the size – a fraction of a carat) is secured on parallel planes to the base of the electronic part of the device. To ensure high heat-transfer capacity, when brazing the gaps between the diamond crystal and the metallic part should be minimum.

In thermistors, a diamond crystal with a size of $0.5–1 \cdot 10^{-3}$ m, ground and polished on two opposite parallel faces, is bonded to wire electrodes (by brazing). Brazing of electrodes can be carried out to one plane of the diamond crystal at a specific distance from each other.

Diamond sound pickups

A combined diamond-metallic needle of a sound pickup is bonded to the metallic casing in sound reproduction systems.

This type of needle is produces using the following procedure. A diamond grain is placed in an end hole of a vertical steel or titanium cylinder with a diameter 1.5 mm, height $8 \cdot 10^{-3}$ m. The brazing alloy is placed on the top and vacuum brazing is carried out. Subsequently, the diamond together with the holder are machined to the required shape and dimensions. In all examined cases, the diamond is in contact with the metal and the strength of this contact determines the quality and properties of the produced diamond components.

Diamond anvils in high-pressure apparatus

The authors of Ref. 265 developed a method of securing diamond anvils in high-pressure apparatus by brazing pre-metallised diamond to metallic supports. This method is used instead of the mechanical (fixing with screws, rolling into a soft ring or a socket of the support) and adhesion methods.

The anvils are produced using a diamond crystal with special faceting, Fig. 89, and multilayer metallising, where the lower layer is made of a carbide-forming metal and the upper layer is made of gold.

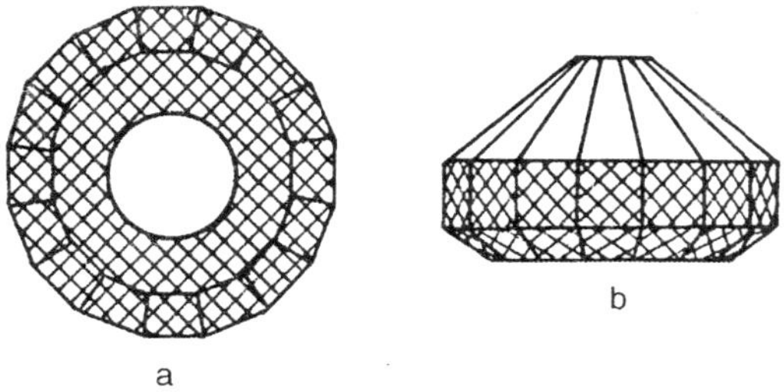

Fig.89 Geometry of metallising diamond anvils: a) view from the side of the base; b) side view.

The metal coates the entire base of the anvil, with the exception of a circular area in the centre, aciting as an optical window.

The following brazing technology is used: the surface of the support or plunger coated with a Pb–Sn brazing alloy, and the residue of the molten brazing alloy is removed to leave only a thin layer. Subsequently, the anvil is coated with the flux and press to the support using a spring device. The device together with the support and the anvil are heated to a temperature of approximately 200°C, which is higher than the melting point of the brazing alloy, and are then left to cool down.

In this case, the aim of metallising is not to increase the strength of the brazed joint but to prevent the excess of the brazing alloy to spread in an irregular manner over the surface of the support and maintain it in the form of a belt at the periphery of the anvil.

In Ref. 266, the authors reviewed the application of brazing alloys of natural and synthetic diamond in electronics, optics, surgery and materials science, based on high hardness, strength, thermal conductivity, the wide range of transparence and other unique properties (see Fig. 90–92).

In electronics, diamond substrates are used for the manufacture of powerful and reliable devices, for example, laser diodes, used as retranslators in fibre optics communications systems.

In optics, diamond is a promising material for producing windows

Fig. 90 Diamond heat sinks brazed into copper cylinders.

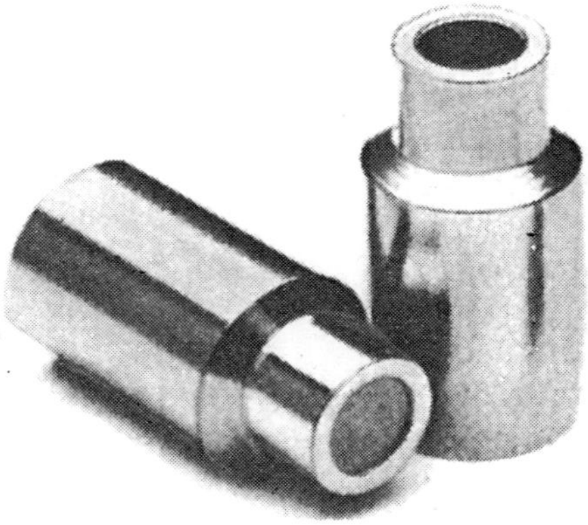

Fig. 91 Brazed diamond windows.

Fig. 92 Brazed diamond blade.

for infrared radiation resistant to corrosion and abrasive effects.

In surgery, brazed diamond blades make it possible to carry out finer and more accurate cuts with minimum damage to the vessels, ensuring rapid healing of incisions after operations such as the removal of cataracts, etc.

4.2 Brazing alloys and fluxes

The brazing alloys used for brazing can be subdivided into two groups. The first group includes metals or alloys that do not interact chemically with carbon, do not form adhesion bonds with the diamond surface and do not wet the diamond. When using brazing alloys of this group, diamond is secured only by mechanical contact. The penetration of the brazing alloy into the brazing gap can be achieved only if the metallic base of the component or the powder metallic frame of the matrix of the tool are efficiently wetted by the brazing alloy.

When using these brazing alloys, to ensure chemical bonding with diamond, the latter is coated with a metallic film – an interlayer which, forming a strong bond with the diamond, has metallic properties, is efficiently wetted and bonded with the metallic brazing alloy (see Chapter 1).

The second group includes the blazing alloys with sufficiently high chemical affinity for carbon obtaining carbide-former elements and forming strong adhesion bonds with diamond. The content of the chemically active element, one or several, is usually not high, 1–10%. In many cases, the brazing alloys of the first group are used as a base into which the carbide-forming elements are then introduced.

Various brazing alloys will now be examined. One of the most important characteristics of the brazing alloys for brazing diamond is the melting point which should not exceed the temperature of rapid graphitisation of the diamonds (~1250 °C). Copper–silver brazing alloys with different alloying elements are used widely. These brazing alloys have relatively low melting points (600–800 °C), are sufficiently ductile, and have good fluidity.

The melting point of the Cu–72% Ag eutectic alloy is 780 °C. The addition of metals such as zinc, cadmium, and tin decreases the melting point of the brazing alloy and this has a positive effect on the performance of diamond tools.

Several compositions of the alloys containing copper and silver and recommended for brazing diamond are given below: 72% Ag, the rest is Cu, t_m = 780 °C; 70% Ag, 25–26 % Cu, 4–5% Zn, t_m = 750 °C; 40% Ag, 21% Zn, 20% Cd, rest – Cu, t_m = 620 °C;[267] 58–59% Au, 1–2% Ag, 0.9–1.1% Zn, 2.5–2.7% Cd, 4–4.2% In, 0.004–0.006% B, Cu – rest;[268] 20–30% Zn, 5–10% Mn, 5–8% Ag, 2–5% Sn, Cu – rest; 20–30% Zn, 5–10% Mn, 5–8% Ag, 2–5% Sn, 1–3% Ni, Cu – rest, t_m = 820 °C.[269]

Several compositions of silver-free brazing alloys, recommended for brazing diamonds, have been published in Ref. 270–272. Low-temperature brazing alloys based on Zn for brazing diamond were discussed in Ref. 270 (Table 24).

A high-temperature brazing alloy based on nickel and containing borides of chromium, molybdenum and titanium in an amount of 5–25% was proposed by the authors of Ref. 271 for brazing cermet diamond-bearing dies to the steel body of the tool.

The authors of Ref. 272 examined an alloy for producing diamond tools, for example, crowns and also saws and cutters, based on the carbides of metals and alloying elements with the hardness of the matrix of HB 400–600, containing 3–5% of carbide or carbides of tungsten, titanium, chromium, vanadium, 35–60% nickel, 0.1–10% tin, 0.1–5% Zn and 0.1–5% Al.

Brazing alloys and binders based on bronze are used widely (Cu with 10–20% Sn).[273]

The absence in these compositions of the brazing alloys of elements with adhesion activity in relation to the diamond does not make it possible to ensure sufficiently high adhesion to the diamond and the adhesion strength is similar to the strength obtained by surrounding the grain by the metal and is sufficiently high only for some types of tools.

The brazing alloys of the second group, characterised by chemical

Table 24 Zinc-based brazing alloys

Composition, %	$t_{melting}$, °C	$t_{brazing}$, °C
60Zn–20Ge–20Al	360	400
75Zn–20Ge–5Al	420	470
75Zn–5Ge–20Al	450	500
90Zn–5Ge–5Al	370	420
62Zn–8Ge–30Al	470	510
79Zn–10Ge–9Al–1Ni–1Ag	355	400
85Zn–8Ge–5Al–2Ag	360	400
82Zn–10Ge–6Al–1Ni–1Ag	420	460
83Zn–5Ge–10Al–1Ni–1Ag	430	480
60Zn–20Ge–15Al–1Ni–1Ag–3Si	440	490
83Zn–7Ge–5Al–1Ag–4Si	410	460
60Zn–30Ge–5Al–2Ni–3Si	510	560
55Zn–10Ge–33Al–1Ni–1Ag	510	560

affinity for carbon, are highly promising. In most cases, these brazing alloys usually consist of a base represented by a metal with a relatively low melting point of 300–1100°C (copper, silver, gold, tin, and others, and also their alloys, with eutectic composition in some cases), chemically inert in relation to the diamond surface, and the active additions of the elements with high affinity for carbon. These are mainly transitional elements: titanium, chromium, tantalum, niobium, vanadium, zirconium, and others.

To develop brazing alloys of this type, it is important to have data on the wetting of the diamond surface by binary or multicomponent alloys in relation to concentration – the wetting diagram, and the data on the strength of the diamond–metal contact (see Chapters 1 and 3). These data can be used in selecting the alloys with the required capillary and adhesion properties. The brazing alloys can be represented by the alloys of the following systems: Sn–Ti, Ag–Ti, Cu–Cr, Cu–Ga–Cr, Cu–Sn–Ti, Cu–Ga–Ti, Cu–Ag–Ti, Au–Ta.

The content of the active component should be restricted because increasing concentration of this component often increases the degree of embrittlement of the alloys thus increasing the stresses in the brazed joint.[45] The true contact strength also decreases when the content of the chemically active component is increased above the optimum level.[47,48] The authors of Ref. 7 used brazing alloys with very small additions of an active metal (fractions of a percent).

The authors of Ref. 274 proposed the brazing alloy of the following composition: 2.85–14.6% Ti, 54.43–68.84% Ag, 21.19–26.77% Cu, 1.44–11.08% Ni. The brazing temperature is 1000–1150°C, brazing is carried out in vacuum.

Brazing alloys based on the eutectic composition Cu–72% Ag with additions of 1–10% Ti have been discussed quite extensively in the literature.[275–277] These brazing alloys are either melted together with the base or deposited on diamond by various methods; this is followed by brazing with a Cu–Ag brazing alloys. Brazing temperature in this case is 850–900°C.

The alloys on the Cu–Sn base, with additions of titanium, have been described and investigated quite extensively.[45,61] According to Ref. 45, the optimum ratio of the component from the viewpoint of the efficiency of the produced diamond tools is the composition containing 20% Sn, 10% Ti and Cu, brazing temperature 960 °C.

Other alloys, designed for brazing diamond, consist of a Cu–Ga base with chromium additions.[278]

The authors of Ref. 279 proposed a brazing alloys based on Al containing from 8 to 18% silicon, with brazing carried out in vacuum or an inert gas at 570–670°C. The alloy is ductile and can be easily processed to produce wire by drawing or foil by rolling. The main advantage of the brazing alloy is, according to the authors, it's low brazing temperature in contrast to the copper and nickel brazing alloys.

The high-temperature brazing alloys based on nickel and Cu with the compositions 70% Ni, 20% Cr, 10% B, Si, P; 25–45% Cr, 25–55% Cu, 10–30% Mn, 2–10% Sn, Cd, Pb, Bi, 1–7% Ti, 0.1–1.5% P have a melting point of ~1200°C and can be used in vacuum brazing of diamond according to Ref. 280,281.

The authors of Ref. 282 proposed the composition of a nickel-based brazing alloy with additions of Sn (7–10%) and Cr (30–35%) for brazing polycrystals of synthetic diamond in dies.

The brazing alloys included in the first group can be used for producing tools in air under a flux. This is determined by their low oxidation susceptibility. It is preferred to use a brazing alloy with a relatively low melting point (up to 800 °C). Taking into account in these cases the low strength of contact of the metal with diamond, the latter should be inserted no more than 2/3 of its length into the matrix of the tool.

Brazing of diamond with brazing alloys included in the second group can be carried out only in high vacuum or a shielding medium. This is explained by the high oxidation susceptibility of the majority of metals that are active in relation to carbon. In this case, the high contact strength makes it possible to braze diamonds without insertion into the body of the tool holder. The application of the brazing alloys of the second group results in a bonding strength of up to 50–200 MPa.

However, the examined brazing alloys of the second group are often brittle because of the presence of a large amount of elements included in their composition, and the formation of brittle intermetallic compounds.[2,45,283,284] Large quantities of the active additions also result in excessive increase of the thickness of the carbide interlayer at the diamond–metal interface,[275,277,285] thus increasing the stresses and influencing the strength of the diamond–metal joint.

When using the brazing alloys of the first group, brazing can be carried out in air. Relatively low temperatures in heating (to 700 – 800°C), and the short duration of the brazing process make it possible to retain the diamond grain without burning it. However, the metallic component oxidises during brazing and, consequently, it is not wetted by the brazing alloy. Fluxes must be used in such cases. They are subject to usual requirements: a flux must melt and interact with the oxides on the surface of the metal prior to melting of the brazing alloy, must wet moderately the brazed materials – metal and diamond, support spreading of the brazing alloy on the metallic brazed surface, retained the chemical composition and capillary properties during the brazing cycle; the flux must partially prevent oxidation (combustion) of diamond.

The majority of salt melts form small contact wetting angles on diamond[286,287] so that the flux during melted penetrates into the gaps of the fitting sockets of the diamond–metallic component, comes into contact with the metal surface, cleans and deoxidises it. The metallic brazing alloy wets efficiently the metallic component and, consequently, regardless of the fact that it does not wet the diamond surface, penetrates into the brazing gap and comes into contact with the diamond surface. Being a lighter melt, the flux is displaced to the upper part of the fitting hole if the latter is in the vertical position.

The flux in brazing is selected on the basis of the material of the brazing alloy and the metallic part of the component. There are fluxes for high-temperature brazing (temperature range of their activity is 450-1300°C) and for low-temperature brazing (150–400°C).[287]

In most cases, brazing of diamond and other superhard materials in toolmaking is carried out using high-temperature brazing fluxes because the melting point of the brazing alloys is usually higher than

450 °C. The following fluxes are used: sodium tetraborate $Na_2B_4O_7$ 100%, working temperature 800–1150 °C; No.209; potassium fluoride (dehydrated) 33–37%, potassium fluoroborate 21–25% boric anhydride 33–37% (600–850 °C); No. 284: potassium fluoride (dehydrated) 33–37%, potassium fluoroborate 40–44%, boric anhydride 23–25% (500–850 °C).

The fluxes are introduced into the brazing zone by sprinkling them onto the assembled structure to be brazed (metallic component, brazing alloy, diamond tool). It is also possible to use a pool with a molten brazing alloy, tubular brazing alloys in which the internal part is filled with the required flux, and other methods.

4.3 Heating methods and equipment

The brazing of diamond can be carried out in the same equipment as brazing of metals, especially electric resistance furnaces, induction heating equipment, and gas torches or plasma heat sources can also be used.

When using adhesion-active brazing alloys, containing chemically active elements – titanium, chromium, vanadium and other elements, the brazing of diamond should be carried out in high vacuum (10^{-3}–10^{-4} Pa). This is the most efficient method of securing the diamond.

Vacuum furnaces of various design are used.

Industry in Russia and Ukraine produces several types of vacuum furnaces suitable for brazing the diamonds and producing brazed components contain diamonds. They include[288] vacuum chamber-type electric furnaces of the SNV 1.3.1/16 I1 (201) type, intended for operation at temperatures of up to 1600–1800 °C and a vacuum of 1×10^{-3} Pa. The dimensions of the working space are 0.1×0.3×0.1 m, power 22–40 kW.

The shaft-type electric furnaces OKB-8085 (8086) can be used for heating to 1100–1600 °C in a vacuum of 2.7×10^{-3} Pa. The dimensions of the working space of the furnaces are 0.4×0.6 m, power 127 kW. The SGV 2.4.2/15I2 two-bell electric furnaces can be used for heating to 1500 °C, in a vacuum of 5×10^{-3} Pa. The dimensions of the working space are 0.3×0.4 m, power 25 kW.

In practice, the total power of the furnaces is not utilised completely. The temperature required in brazing diamonds is 900–1200 °C.

4.4 Organisations of the technological section for adhesion brazing and metallising diamonds under industrial conditions

When using adhesion-active brazing alloys and fluxless vacuum brazing of diamonds, it is necessary to fulfil stringent requirements for vacuum hygiene, purity and culture of production. These stringent conditions can also be fulfilled under the plant conditions. The vacuum furnace (or several furnaces) must be positioned in a separate clean, light room. Rough vacuum pumps must be situated and assembled also separately (in a corridor, basement) and must be connected by pipelines directly with the high-vacuum equipment. The pipelines are usually produced from stainless steel pipes, 80–100 mm in diameter. The length of the pipelines can reach several tens of metres and, in this case, the pipeline is used as the container of the rough vacuum pump. The pipeline is connected to the rough vacuum prior using a vacuum valve controlled remotely from the operating area of the operator. The exhaust of the pump is discharged into the atmosphere using a special pipeline.

The vacuum systems must be maintained clean. Contamination or dust are not permitted. The non-working apparatus must be under vacuum. The recommended operating regime of the furnaces is to load the component for brazing at the end of the working day or during the evening shifts, preliminary pumping, holding under vacuum during the night, and the working cycle heating is carried out the following day. Preparation of components for brazing and assembly of the components should be carried out in a separate section adjacent to the room where the vacuum furnaces are installed. The working table for the assembly of the components is fitted with an extraction system. An extraction box can be used.

The surfaces of the components to be brazed must be clean and, therefore, special equipment is provided for rinsing, cleaning and drying components (diamonds, in particular).

Assembly is carried out using various tools – pincers,clamps. Brazing equipment should provide for a group technological process in which a large number of components is loaded into the furnace during a single cycle (depending on the dimensions of the furnace space and components), to 10–100 or more pieces. All components of the equipment-holders, containers, crucibles, etc. are subjected to preliminary heating at temperatures higher than the process temperature.

The personel of the section consists of an engineer and an operator of the vacuum furnace. The shaping of components and assembly for brazing are carried out by a technician or a worker. The total number of personel is determined on the basis of the volume of production of components, their dimensions and shape, the degree of automation of the process and other factors.

4.5 Technology of brazing diamond materials

The technological processes of adhesion brazing diamond materials can be divided into single-stage using adhesion-active alloys (in melting, these alloys wet the diamond surface and form strong interphase bonds with carbon atoms) and two-stage processes in which the diamond is coated in advance with a metallic film (the diamond is metallised) and the brazing process can then be carried out using a conventional brazing alloy used for joining metals.

The strength of the brazed joint in the first case is determined by the interaction of the brazing alloy directly with the diamond surface, and in the second case it depends on the bonding strength of the metallised film of the coating with the diamond surface.

Metallising of the diamond is carried out using one of these methods. From the viewpoint of the maximum contact strength, the most efficient methods are those accompanied simultaneously by the heat treatment of the diamond with the coating under the optimum conditions where chemical bonds of carbon with the adhesion-active metal form and the diffusion softening of the contact has no effect.

Metallising of the diamond by the contact-reaction method by annealing it in a mixture with a powder metallised under specific conditions results in sufficiently high bonding strength of the coating film with the diamond surface reaching 150–200 MPa in separation. The highest values of the bonding strength, obtained in single-stage brazing in the case of titanium alloys were 300–350 MPa. It is important to select correctly the metallic component.

It was shown previously that the presence of the metals of the iron group in the interphase region results in the graphitisation of diamond during heating the composition and greatly reduces the strength of the brazed joint. In this case, it is efficient to use refractory metals – molybdenum, tungsten, niobium, tantalum, etc, but the application of these metals as structural materials is limited.

When using steel dies or holders in diamond tools, it is necessary to take measures to prevent the penetration of the atoms of iron, nickel and manganese into the interphase region. This is achieved by increasing the thickness of the metallising coating using two-stage brazing technology (reducing the 'permeability' of the coating); using interlayers or inserts of molybdenum, for example those brazed into the fitting socket and insulating the diamond from the steel surface; saturation of the surface of the steel component in the joint area by the element bonding the metal of the iron group into strong chemical compounds. Case hardening, siliconising, chrome plating, boronizing and other processes can be used here.

Several brazing processes will now be examined. Brazing of diamond crystals when producing single crystal tools is often carried out using copper–silver brazing alloys. In most cases, brazing is carried out in air under a flux using induction heating. To secure diamond in the holder, it is mechanically caulked in the fitting hole.

This technology cannot ensure sufficiently high strength of the diamond-metallic contact and, consequently, the diamond must be inserted into the holder to 2/3 of its length. After brazing, the diamond element is mechanically cleaned to 1/3 of its length to remove the remnants of the brazing alloy and the flux and it is then processed in special faceting machines to the required shaped and dimensions. This method is used to produce lathe cutters, cutters for drilling continuous holes (Fig. 86a) and separating cutters (Fig. 86b), etc.

When producing diamond cutters for profiling worm grinding wheels, Fig.86d, it is necessary to use a different technology. Diamond crystals are pressed together with the iron or copper–iron powder into the hole in the holders. This is followed by vacuum infiltration of the skeleton by molten copper. In this case, the strength of the diamond–metal contact is also low and the diamond is secured in the socket only by the mechanical effect of the binder.

High-strength diamond–metal contact can be produced using the following procedure.

1. The use of the alloy (brazing alloy) containing additions of chromium and titanium, and direct brazing of the grains and crystals of diamond using this alloy. The base can be represented by the Cu–18–19% Ga alloy, the addition of chromium is 0.4–1.0%, titanium addition 0.7–1.0%. Brazing is carried out in a vacuum of $(1–6)\times10^{-3}$ Pa at 1000°C. The strength of the brazed joint is 100 and 300 MPa, respectively.

2. Metallising of the diamond with chromium or molybdenum followed by brazing with any brazing alloy with sufficiently high ductility. The optimum conditions for depositing the chromium coating are 1000–1030 °C, for the molybdenum coating 950–1000 °C, 60 minutes. The maximum strength in the first case reaches 150–160 MPa, in the second case 120–130 MPa.

When producing combined diamond–metallic needles, the diamond grains are coated with a chromium layer 1.5–2.0 μm thick (by the contact-reaction method) and are then brazed in the hole of the steel holder using a ductile copper-based brazing alloy. Both processes (metallising and brazing) are carried out in vacuum. High strength of the diamond-metallic contact (to 150 MPa) makes it possible to produce high-quality components.

If the coating on the diamond is not thick (to 1 μm), and brazing of the diamond to the steel holder is carried out using the brazing alloys with the melting point higher than 900 °C, graphitisation of the diamond and softening of the contact can take place. In this case, it is necessary to use interlayers (molybdenum, tungsten) with a thickness of several fractions of a millimetre positioned between the steel holder and diamond, thus preventing the penetration of iron atoms to the diamond surface during brazing and, consequently, graphitisation of the diamond.[289] In industry, the extent of application of these interlayers in large-series production is still small.

Preliminary chrome plating of the steel holder also makes it possible to produce sufficient strength of bonding the diamond with the metal (80–100 MPa).[290]

There is a method of brazing diamond by preliminary coating with zirconium followed by brazing with silver. This method is used in producing straightening cutters, Fig. 86c. It should be noted that the application of zirconium is complicated by its very high oxidation susceptibility. The oxidised surface of the zirconium coating loses its capacity for wetting by the metal of the brazing alloy.

Fine diamond grains can be secured in tools using the following procedure. For example, when producing grinding wheels based on a binder consisting of the Al–Si–Sn alloy at temperatures of 960–1000 °C the Cu–Ti–Ni or Cu–Sn–Ti coating is deposited on the diamonds by the liquid phase sintering method. The thickness of the deposited coating is 5–10 μm, the bonding strength with diamond is 40–50 MPa. During fabrication of the tools, the binder melts (650–700 °C) and interacts with the metallic coating. The results of examination of the interactions show that there is no sharp interface between the binder and the Cu–Sn–Ti alloy and the components of the alloys penetrate into each other. When using the Cu–Ti–Ni coating, there is a sharp interface between the coating and the binder with a considerably smaller zone of penetration in the depth. Analysis of the experiment results shows that the intensity of interaction increases with decreasing difference between the chemical composition of the alloy of the coating and the binder.[291] Comparison of the results of tests on the diamond wheels, produced from metallised diamond powders, with data obtained in examining the interaction of the metal of the coating with the metal of the binder shows that the increase of the durability of the diamond wheels is associated with stronger bonding of the diamond grains in the binder as a result of interaction between the metal of the coating and the binder.

Similar investigations were carried out in Ref. 292 when developing highly efficient tools for processing glass and other brittle materials. A Ni–Mn–Sn–Ti coating was deposited on diamond in vacuum from the liquid phase at a temperature of 900 °C. Bonding of the coating to the diamond surface was ensured as result of the formation of manganese carbide Mn_7C_3. Subsequently, the standard procedure was used to press wheels from the metallised diamond and the binder (Cu–20% Sn) followed by sintering at 750 °C.

When the temperature during sintering was increased, molten tin interacted with the copper with the formation of a solid–liquid melt of a complicated composition which, interacting with the metal of the coating, formed a diffusion layer approximately 0.05–0.06 mm thick, spreading from the initial interface to the side of the binder indicating the diffusion of the coating metal into the binder. The diffusion zone is characterised by a higher content of nickel and manganese (diffusion penetration of titanium into the binder was not detected). This indicates the formation of an adhesion bond between the metallised diamond with the metallic binder and efficient securing of the diamond grain, as confirmed by the higher durability and productivity of the tools made of diamond with the Ni–Mn–Sn–Ti coating.

At the same time, the experimental results show that metallising of the diamond surface has almost no influence on the bonding strength of diamond with an organic (resinoid) binder,[293] although an increase of the strength of the diamond grains itself as a result of metallising has a positive effect on the operating efficiency of the tool. However, even under the conditions of efficient bonding of the surface of the diamond grains with the binder, the grains can be quite easily chipped out together with the binding material because of the low mechanical properties of the binder.

Therefore, to increase the strength of bonding of the diamond in the tool using the organic binder, it is efficient to use diamond grains in the form of the so-called aggregates which enable a considerably larger volume of the binder material to participate in securing the diamond. Using the method proposed in Ref. 105, the aggregate grains are particles of branched form consisting of two or more elementary grains of the abrasive material brazed together and coated on the side of the surface by the adhesion-active metallic alloy (Fig.93). One should note the importance of producing aggregates of highly branched form. A number of patents by foreign companies[294-296] for the preparation of the diamond aggregates do not satisfy this requirement. For example, there are methods of producing aggregated diamond grains in which a mixture of diamond and metallic powders, especially copper, nickel, cobalt or their alloys, is pressed. The metallic and diamond powders are taken in ratios at which the volume of the pores between the diamond particles is filled. The intensity of interaction of the selective binder metals with the diamond is low. The produced cake, having the form of a dense abrasive solid, is cooled down, crushed, thus producing aggregates in the form of fragments of the metal with the embedded

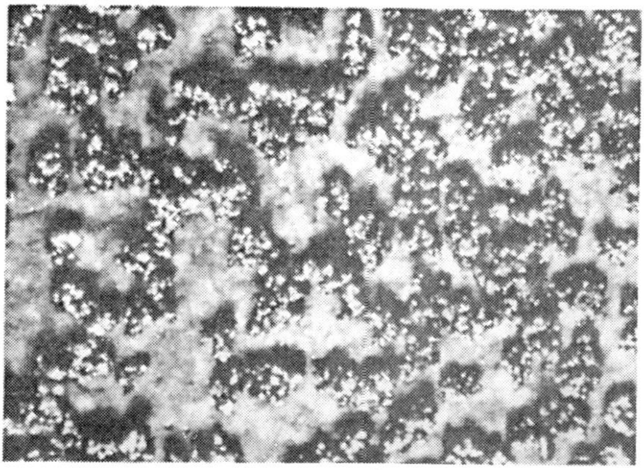

Fig.93 Aggregate diamond grains.

abrasive grains; these aggregates are of no branched form.

The highly branched aggregates are produced by the methods proposed in Ref. 105 by liquid-phase sintering a freely distributed mixture of diamond and metallic powders taken at specific ratios, using, as the metallic component, the adhesion-active (in relation to the diamond/abrasive) powder alloy containing carbide-forming metals titanium, chromium, zirconium, resulting in very low (almost zero) contact wetting angles of the diamond. Because of the high capillary properties, this melts spreads efficiently over the surface of the grains, bonds them together and the high initial porosity of the composite makes it possible to produce aggregates with a developed branched structure. Combined aggregation of the diamond and other abrasive powders results in aggregates consisting of the grains of diamond and the abrasive. The role of the second abrasive as a support for the diamond grains becomes also important in this case.[1] Therefore, the second abrasive takes part in the grinding operation and the range of application of tools of this type can be greatly widened.

The technology of production of the grinding tools of various types and dimensions using the organic binder with aggregated metallised diamonds does not differ greatly from the technology of producing these tools from uncoated diamond. It is only important to take into account the fact that the part of the volume of the charge, for example, of the grinding wheel is occupied by the metal of the coating. This amount for different degrees of metallising varies from 8 to 15% of the total volume of the diamond-bearing part of the tool. The volume of the diamond-free part of the metallic composition must be reduced by this value. The results of special investigations have shown that the variation of the amount of the binder in this range has no negative effect on the strength characteristics of the diamond-bearing layer of the tool.

On the other hand, the application of the branched aggregates of the diamond grains, which have penetrated into the binder and produced a 'sleleton', increases the strength characteristics of the tool composition.

Depending on the abrasive used, the coating and their quantitative ratio, the tools produced on their basis can be used for the processing of high-strength brittle materials, brittle materials together with ductile materials (for example, a hard alloy with steel), various steels, titanium alloys, and new structural materials (see Fig.94).

The results of a large number of laboratory and industrial tests have shown that the highest abrasive properties are observed for the tools produced from the jointly aggregated metallised diamond-abrasive powders.[1,297]

Impregnation is another method used widely in producing diamond tools, in addition to sintering. In this case, wetting of the solid body by the liquid is the main process ensuring the penetration of the liquid into the pores and capillaries of the solid. At a contact wetting angle smaller than 90° ($\cos \Theta > 0$) spontaneous impregnation takes place, and $\Theta > 90°$ ($\cos \Theta < 0$) impregnation does not occur. Intensive impregnation requires a relatively low value of the contact wetting angle (40–10° or lower), i.e. it is necessary to ensure a high degree of wetting of the solid phase by the liquid.

Owing to the fact that the majority of metals with a melting point of up to 1000°C (the degree of graphitisation of diamond is still low) do not wet diamond in vacuum and, consequently, they cannot impregnate the diamond powders, metallic coatings can be deposited on the diamond using the method described previously. For example, in Ref. 298, the method of reaction deposition was used to deposit on the dia-

Fig.94 Grinding wheels for treatment of hard alloys.

mond molybdenum coatings of various thickness and this was followed by examining the impregnation of the layer of the metallised diamond particles by the metallic melt. The impregnating material in this case was a copper-based alloy alloyed with nickel and manganese, with a melting point of 980 ± 20°C. The impregnation kinetics[299] was determined from the increase of the depth of the layer of the molybdenum-coated diamond powders of different grain size (film thickness 0.5 μm) impregnated with the copper–nickel–manganese alloy. Their initial porosity was 47–55%, depending on dispersion. The results, Fig. 95, are in agreement with the literature data which indicate that increasing particle size of the solid phase from which the porous solid is produced and, consequently, increasing diameter of the pore channels increases the rate of impregnation. Theoretically, this is associated with the fact that the constant of the impregnation rate in the equation, describing this process, is directly proportional to the radius of the capillary. Satisfactorily straightening of the experimental kinetic curves in the square of the height of the impregnated layer-time coordinates indicates that the movement of the metallic liquid in the layer of the metallised diamond grains is governed by the parabolic law, i.e. the square law of the dependence of the depth of impregnation on time is also confirmed satisfactorily in this case. The investigations represent a basis for developing the technology of producing diamond tools with a metallic binder by the impregnation method using adhesion-active coatings. The diamond-bearing layer of these tools is characterised by increased concentration of adhesion-secured diamonds, high mechanical strength, and is used efficiently in the tools for processing brittle material (tubular drills, hence, cutters, straightening rollers, grinding disks). The technology of fabrication of these tools is characterised by strong bonding of the diamond-bearing layer with the body of the tools and

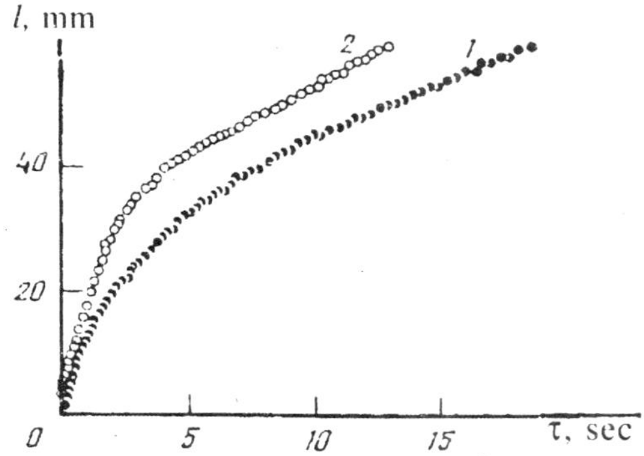

Fig.95 Kinetics of impregnation of molybdenum-metallised diamond powders with a copper–nickel–manganese alloy: 1) AC15 63/50; 2) AC15 400/315.

makes it possible to produce components of greatly differing shapes and dimensions (Fig. 96).

Another variety of the diamond-abrasive tool are disk saws for cutting rock whose durability and productivity can also be greatly improved by improving the efficiency of securing the diamond grains in the metallic matrix. The experimental investigations show[300,301] that the deposition of appropriate adhesion coatings on the diamond or the use of adhesion-active binders greatly improves the quality of the tools. In Ref. 302, the binders were selected taking into account of the physical-mechanical properties of the existing binders (based on copper–tin, copper-iron, hard alloys and other bases) for these purposes because they have the required hardness, wear resistance and self-sharpening capacity. The adhesion properties of these binders were improved by adding to them elements increasing the adhesion strength of the diamond-metallic contact, Ti, Cr. Rock cutting tools made on the basis of these binders showed higher efficiency in processing granite and marble.

In Ref. 303, the fabrication of tools by the impregnation of sintering method was carried out using diamond with tungsten, tantalum, molybdenum, and niobium coatings 1–30 μm thick on which copper was additionally deposited as the protective layer. The refractory metals were deposited by the chemical method from the vapour phase in a graphite reactor into which argon was supplied at a specific pressure and for example, tungsten fluoride for depositing a tungsten coating on the diamond.

The segments for saws with the matrix of the Cu–Sn alloy, containing up to 20% of tungsten carbide grains, and produced on the basis of these metallised diamonds showed a lower diamond consumption when cutting siliceous limestone by almost 30% in comparison with uncoated diamonds.

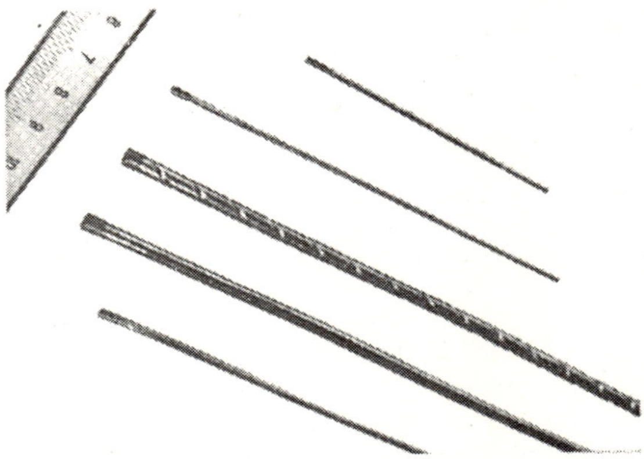

Fig.96 Drills of different standard dimensions for treatment of nonmetallic materials.

The authors have also reported that the Diamet diamond-bearing material produced by them and consisting of tungsten-coated diamond grains of two sizes (around 500 and 60 μm), impregnated with the copper–manganese–titanium alloy in vacuum at 1050 °C, was used efficiently for cutting granite. This material can also be used, according to the authors, in fabrication of drills and drill columns.

Brazing of diamond in drilling bits

The diamond drilling bits for drilling hard rock are usually produced by impregnation of the powder composition, consisting of diamond and a hard alloy, with a metallic alloy-binder. The binders are presented by copper or alloys of copper and nickel. These binders are not adhesion-active and do not wet diamond grains. Therefore, in this case, the grains are secured only by the mechanical encirclement by the binder metal.

To determine the bonding strength of the metallised and uncoated diamond grains with the matrix of the drilling columns, investigations were carried out to simulate the industrial process of production of drilling bits. For this purpose, a powder of a hard alloy was pressed into a steel sleeve together with the binder. The specimen was placed on the top of a diamond crystal (1–1.5 carat) which was placed in a small depression of the graphite matrix. In heating in hydrogen, copper and copper-based alloys were impregnated. The impregnating material bonded to the matrix with the diamond face into a monolith. Impregnation was followed by mechanical tests of the diamond-metallic contact in shear at room temperature. The test results show that in the case of impregnation of the non-metallised diamond with copper or a copper-nickel only, the shear strength of the contact was 1–2 MPa. The application of a number of developed coatings, consisting of one or more refractory carbide-forming metals (chromium, molybdenum, tungsten), together with relatively low-melting metals resulted in relatively high strength of the contact (90–100 MPa).

The results of these investigations were used to select the composition and develop a technology of depositing a multi-layer coating on the diamond. This coating is strongly bonded to the diamond, prevents diffusion softening of the binder and graphitisation of the diamond in high-temperature impregnation and operation of the tool. In addition, this coating is sufficiently wetted by the metals forming the base of the binder of the bits.

The application of adhesion-active binders, e.g. copper with small additions of Cr, ensures, because of high adhesion of the alloy to the diamond, strong bonding of the grains in the matrix of the bit in impregnation of the composite produced at 1100–1150 °C in high vacuum. In this case, there was almost no dropping out of the diamond grains.

The industrial and laboratory (stand) tests of standard drilling bits and bits fitted with diamond with a multi-layer coating and also saturated with the adhesion-active binder, in drilling granite of grade 9 and 10 as regards drillability showed that the working life increased on average by 25–30% and the consumption of diamond decreased by almost the same value.

The positive effect of metallisation of the diamonds on the efficiency of the drill bits was also reported in Ref. 304. The two-layer titanium-nickel coating was deposited on the diamond by deposition in vacuum in crossed electric and magnetic fields.

Tests of single-layer drilling bits of the 01A3 and 01A4 types, carried out in geological surveying wells of 9.0 and 9.5 drillability categories showed that the durability of the the drilling bits increased by almost 35% and the specific consumption of diamond was reduced by almost 20%, with a small decrease of the mechanical drilling rate.

Brazed diamond thermistors

In Ref. 305 brazing of the electrodes of diamond thermistors was carried out using gold-tantalum (1% Ta) and gold–tantalum–aluminium (9% Ta and 1% Al) alloys. The alloy in the form of sheet 0.02 mm thick, with an area of 0.5 mm^2 was placed on diamond and compressed with tungsten wire, Fig. 97. Brazing was carried out in vacuum at 1050°C. This process is not used for synthetic diamond thermistors because metallic inclusions in diamond at temperatures above 950°C have a detrimental effect on its electrical and physical properties.

In Ref. 306, electrical contacts were produced using a Cu–Ag–Ti alloy with a melting point of 900°C. Brazing was carried out in a vacuum of 10^{-3} MPa for 1–2 min. Adhesion of the current input conductor to diamond in this case was 60 MPa.

The authors of Ref.307 and 308 proposed current-conducting oxidation-resistant Cr–Ni coatings for diamond resulting in a contact strength of up to 70 MPa. Their application in processing diamond crystals made it possible to automate the process of faceting

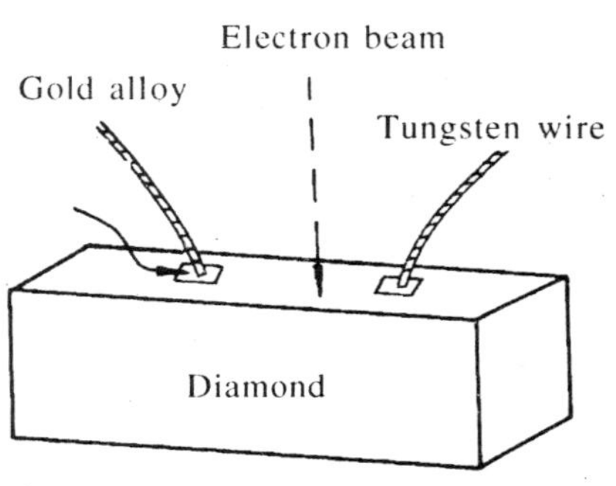

Fig.97 Electron beam welding of contacts with semiconductor diamonds.

semifinished products to diamonds because of retention of the properties of the coating in operation in air by multiple thermocyclic in the temperature range from 20 to 900°C.

CONCLUSIONS

A large number of various brazing alloys and methods of brazing materials have been proposed, including those ensuring very high contact strength between diamond and metal (100–200 MPa). For the individual types of chemically active alloys, the authors of this work have recently obtained contact strength in tensile loading of 300–350 MPa.

The very high values of the strength of the diamond–metal contact on the level of the strength of the metallic or even diamond part of the composite make it possible to build new types and design of diamond–metal components in which the diamond is secured on the plane (a relatively small area, in an open gap, etc) so that for example, in the case of tools, almost the entire volume of the diamond element becomes working, 'cutting', thus increasing the productivity of tools and saving diamond.

Thus, the efficient selection of the adhesion-active alloy or coating material, optimisation of the conditions of formation of the contact and brazing conditions, the orientation of the diamond single crystal during securing-brazing in the direction of high values of the strength of the diamond, and a number of special methods make it possible to increase the strength of the diamond-metal brazed contact and improve of the same time, the efficiency of the diamond–metal component - tools, in particular.

References

1. Yu.V. Naidich, G.A. Kolesnichenko, I.A. Lavrinenko, Ya.F. Motsak, Brazing and metallising of superhard tool materials, Naukova dumka, Kiev (1977).
2. D. Evans, et al, Ind. Diamond Rev., No.9, 306-309 (1977).
3. V.A. Konovalov, Examination of the effect of the strength of diamond bond and wear resistance of metallic binders on the efficiency of diamond-abrasive tools, Dissertation for the title of the candidate of technical sciences, Khar'kov (1974).
4. S. Kh. Zhakhidov, Examination of the strength of bonding the grain in a binder under the effect of temperature and force, Dissertation for the title of the candidate of technical sciences, Moscow (1974).
5. Yu. V.Naidich and G.A. Kolesnischenko, Interaction of metallic melts with the surface of diamond and graphite, Naukova dumka, Kiev (1967).
6. Yu.V. Naidich, Contact phenomena in metallic melts, Naukova dumka, Kiev (1972).
7. P. Scott, et al, J. Mater. Sci., **10**, No.11, 1833-1840 (1975).
8. A.V. Nozhkina and P.P. Otopkov, Almazy, No.5, 1-2 (1971).
9. V.I. Kostikov, Yu.I. Andropov, P.P. Otopkov and A.V. Nozhkina, Wetting of diamond by Ni-Si and Cu-Si alloys, In: Adeziya Rasplavov, Naukova dumka, Kiev (1974).
10. Yu.V. Naidich and G.A. Kolesnischenko, Izv. AN SSSR. Ser. Metally, No.4, 220-230 (1968).
11. Yu.V. Naidich, Thermodynamic and physico-chemical relationships of the phenomena of wetting of solids by liquid metals, In: Wetting and surface properties of melts and solids, Naukova dumka, Kiev (1972).
12. G.A. Meerson and Ya.S. Umanskii, Izv. Sektora Fiz.Khim. Analizov, **22**, 104-110 (1953).
13. G.V. Samsonov, Poroshkovaya Metallurgiya, No.1, 98-106 (1965).
14. Ya.S. Umanskii, Carbides of hard alloys, Metallurigya, Moscow (1947).
15. G.V. Samsonov and Ya.S. Umanskii, Solid joints in refractory metals, Metallurgizdat, Moscow (1957).
16. G. Novotnyi, Crystal chemistry of borides, carbides, nitrides, aluminides and silicides of transmission metals, In: Electronic structure of transition metals and chemistry of their alloys, Metallurgiya, Moscow (1966).
17. A.R. Ubbelode and F.A. Lewis, Graphite and its crystal compounds (translated from English), Mir, Moscow (1965).
18. R. Setton, Bull. Soc. Chim. France, No.10, 1758-1765 (1960).
19. R. Setton, Ibid, No.3, 521-524.
20. F.J. Salzano and S. Aronson, J. Inorg. Chem., **26**, No.8, 1456-1458 (1964).
21. A.R. Ubbelode, Uspekhi Khimii, No.7, 1692-1713 (1938).
22. Ya.S. Umaiskii, Izv. Sektora Fiz.-Khim. Analiza, **16**, 127-130 (1943).
23. T.Ya. Kosolapova, Carbides, Metallurigya, Moscow (1968).
24. L. Bryuer, Thermodynamic stability and the nature of bond in relation to the electronic structure and crystal structure, In: Electronic structure of transition metals and chemistry of their alloys, Metallurgiya, Moscow (1966).
25. C.E. Lowell, J. Amer. Cer. Soc., **50**, No.3, 142-144 (1967).
26. Yu.V. Naidich, V.M. Perevertailo, O.B. Loginova and A.A. Shul'zhenko, Sverkhtverdye Materrialy, No.1, 9-13 (1979).

27. M. Hansen and K. Anderko, Structures of binary alloys, Metallourgizdat (1962).
28. N.K. Adam, Physics and chemistry of surfaces, Gostekhizdat USSR, Moscow and Leningrad (1947).
29. V.K. Semenchenko, Surface phenomena in metals and alloys, Gostekhteorizdat, Moscow (1957).
30. I.A. Sheka, I.S. Chaus and T.T. Midyureva, Gallium, Gostekhizdat USSR, Kiev (1963).
31. F.E. Bowman and R. Cygan, In: Behaviour of sodium in graphite, Liquid-metals handbook, Washington: Atom Energy (1965), pp.152-155.
32. E.F. Anikeev, V.N. Chepelenko, V.I. Kravchuk, et al, Adgeziya Rasplavov i Paika Materialov, No.5, 60-64 (1980).
33. Yu.V. Naidich, V.M. Perevertailo, O.B. Loginova and N.D. Polyanskaya, Sverkhtverdye Materialy, No.4, 17-18 (1985).
34. Yu.V. Naidich and G.A. Kolesnischenko, Poroshkovaya Metallurgiya, No.3, 23-28 (1964).
35. Yu.V. Naidich, V.M. Perevertailo, E.M. Lebovich and A.K. Butylenko, Zhurn. Fiz. Khimii., **51**, No.6, 1483-1485 (1977).
36. Yu.V. Naidich, V.M. Perevertailo and O.B. Loginova, Izv. An SSSR. Ser. Metally, No.4, 37-41 (1979).
37. A.V. Nozhkina and V.I. Kostikov, Wetting of diamonds by melts of transition metals, In: New developments in the theory and practice of production and application of synthetic superhard materials in national economy, Institute of Superhard Material, Academy of Sciences of the Ukrainian SSR, Kiev (1997), pp.166-170.
38. M.A. Maurakh, Trans. Indian Inst. Metals, **14**, 209-225 (1961).
39. J.A. Cahill and A.D. Kirschenbaum, J. Inorg. and Nucl. Chem., **27**, No.1, 73-76 (1965).
40. M. Humenik, et al, Iron Age, **186**, No.19, 171-173 (1960).
41. Yu.V. Naidich and G.A. Kolesnichenko, Poroshkovaya Metallurgiya, No.6, 55-61 (1961).
42. Yu.V. Naidich, Yu.N. Chuvashov and V.S. Zhuravlev, Adgeziya Rasplavov i Paika Materialov, No.1, 50-53 (1976).
43. Yu.N. Chuvashov, V.G. Chuprina and L.P. Obushchak, Ibid, No.7, 59-65 (1981).
44. G.A. Kolesnichenko and Yu.V. Naidich, Almazy, No.3, 8-10 (1970).
45. E.D. Kizikov and I.A. Lavrinenko, Sintet. Almazy, No.6, 21-25 (1973).
46. D. Mortimer, et al, In: Carbon fibres. Their place in modern technology, London (1974), pp.101-104.
47. Yu.V. Naidich, V.P. Umanskii, Yu.N. Chuvashov and I.A. Lavrinenko, Sverkhtverdye Materialy, No.1, 3-6 (1980).
48. Yu.V. Naidich, V.P. Umanskii and I.A. Lavrinenko, Sverkhtverdye Materialy, No.3, 26-29 (1982).
49. V.I. Kostikov and A.N. Varenkov, Interaction of metallic melts with carbon materials, Metallurgiya, Moscow (1981).
50. Yu.V. Naidich, V.M. Perevertailo and E.M. Lebovich, Adgeziya Rasplavov i Paika Materialov, No.1, 28-31 (1976).
51. Yu.V. Naidich and G.A. Kolesnichenko, Examination of the wetting of graphite and diamond by liquid metals, In: Surface phenomena in melts and processes in powder metallurgy, Publishing House of the Academy of Sciences of the Ukrainian SSR, Kiev (1963).
52. Yu.V. Naidich and G.A. Kolesnichenko, Fizika Metallov i Metallovedenie,

18, No.2, 193-197 (1964).
53. Yu.V. Naidich, V.M. Perevertailo and G.M. Nevodnik, Izv. AN SSSR. Ser. Metally, No.2, 87-90 (1972).
54. Yu.V. Naidich, V.M. Perevertailo and G.M. Nevodnik, Izv. AN SSSR. Ser. Metally, No.3, 244-248 (1972).
55. O.B. Loginova, Yu.V. Naidich and V.M. Perevertailo, Adgeziya Rasplavov i Paika Materialov, No.6, 7-10(1980).
56. V.M. Perevertailo, O.B. Loginova, V.G. Delevi, et al, Sverkhtverdye Materialy, No.4, 20-22 (1987).
57. Yu.V. Naidich, V.M. Perevertailo, O.B. Loginova, et al, Sverkhtverdye Materialy, No.2, 10-12 (1989).
58. V.M. Perevertailo, V.G. Delevi, O.B. Loginova, et al, Sverkhtverdye Materialy, No.5, 8-11 (1991).
59. V.M. Perevertailo, O.B. Loginova and Yu.V. Naidich, Adgeziya Rasplavov i Paika Materialov, No.3, 31-33 (1978).
60. Yu.V. Naidich, G.A. Kolesnichenko and N.S. Zyukin, Adgeziya Rasplavov i Paika Materialov, No.10, 57-59 (1982).
61. G.A. Kolesnichenko, N.S. Zyukin and Ya.F. Motsak, Adgeziya Rasplavov i Paika Materialov, No.2, 40-46 (1977).
62. Yu.V. Naidich, Yu.N. Chuvashov, V.S. Zhuravlev, et al, Adgeziya Rasplavov i Paika Materialov, No.5, 8-10 (1980).
63. Yu.V. Naidich, Yu.N. Chuvashov, V.S. Zhuravlev, et al, Zhurn. Fiz. Khimii, No.3, 10-13 (1979).
64. V.M. Perevertailo, V.G. Delevi, O.B. Loginova, et al, Sverkhtverdye Materialy, No.2, 31-33 (1991).
65. V.M. Perevertailo, Sverkhtverdye Materialy, No.4, 7-12 (1991).
66. Yu.N. Chuvashov and V.P. Krasovskii, Adgeziya Rasplavov i Paika Materialov, No.10, 29-31 (1982).
67. Yu.V. Naidich, I.A. Knyazeva, O.B. Loginova, et al, Sverkhtverdye Materialy, No.6, 31-35 (1980).
68. Yu.V. Naidich, A.A. Shul'zhenko, A.V. Andreev, et al, Dokl. AN USSR. Ser. A., No.6, 74-76 (1984).
69. A.S. Vishnevskii and A.V. Lysenko, Kinetics of carbide formation at the interface of the diamond with melts of transition metals, In: Physical chemistry of the condensed phases and their interfaces, Naukova dumka, Kiev (1975).
70. P.P. Otopkov, A.V. Nozhkina, V.I. Kostikov, et al, Vopr. Teorii i Praktiki Almaz. Obrabotki., No.5, 44-51 (1977).
71. Yu.V. Naidich and G.A. Kolesnichenko, Examination of wetting and adhesion of liquid metals to graphite and diamond, In: Surface phenomena in metallurgical processes, Publishing House of Literature in Ferrous and Non-ferrous Metallurgy, Moscow (1963).
72. Yu.V. Naidich and G.A. Kolesnichenko, Poroshkovaya Metallurgiya, No.2, 76-79 (1968).
73. A. Epstein and J.E. Gieb, Electrochem Soc., **106**, No.11, 995 (1959).
74. Yu.V. Naidich and G.A. Kolesnichenko, Examination of contact interaction and wetting of hexagonal and cubic boron nitride by molten metals, Proc. 9th Nat. Conf., Riga (1968), pp.142-146.
75. G.A. Kolesnichenko, Wetting of refractory covalent crystals by metallic melts, In: Wetting and surface properties of melts and solids, Naukova dumka, Kiev (1972).
76. Hauffe K., Reactions in solids and on the surface (translated from Ger-

man), IL, Moscow (1963), Vol.2.
77. V. Seith, Diffusion in metals. The process of exchange of sites, IL, Moscow (1958).
78. G.A. Kolesnichenko, A.S. Vishnevskii, Yu.V. Naidich, et al, Examination of the composition and structure of phases in interaction of adhesion-active melts with diamonds, In: Adgesiya Rasplavov, Naukova dumka, Kiev (1974).
79. S.A. Klevtsur and A.D. Kotok, Metallising of natural diamonds by the thermodiffusion method, In: National Scientific and Technical Conf. on New Methods of Depositing Coatings by Spraying, Voroshilovgrad Engineering Institute, Voroshilovgrad (1975), pp.168-169.
80. I.P. Kushtalova and A.F. Stasyuk, Relationships governing the formation of boundary layers in the diamond-transition metal system, In: Processes of interaction at the phase interface, Institute of Superhard Materials, Academy of Sciences of the Ukrainian SSR, Kiev (1982).
81. A.S. Vishnevskii, A.F. Nikityuk, T.D. Ositinskaya, et al, Interaction of synthetic diamond with melts of transition metals, In: Adgeziya Rasplavov, Naukova dumka, Kiev (1974).
82. V.G. Delevi, I.A. Knyazeva, I.A. Lavrinenko, et al, Poroshkovaya Metallurgiya, No.9, 46-50 (1984).
83. A.M. Varenkov and V.I. Kostikov, Izv. Vuzov. Cher. Metallurgiya, No.3, 138-140 (1975).
84. Yu.V. Naidich, Progr. Surface and Membr. Sci., **14**, 353-484 (1981).
85. E.M. Chistyakov, A.A. Shepel', T.M. Dudan, et al, Tools made of metallised superhard materials, Naukova dumka, Kiev (1982).
86. Yu.V. Naidich, I.A. Lavrinenko and G.P. Volk, Physico-chemical fundamentals and methods of depositing metallic coatings on diamond and development of tools with increased service properties, In: Intl. Seminar: Superhard Materials (Kiev, June 1981), Institute of Superhard Materials, Vol.1, Academy of Sciences of the Ukrainian SSR (1981), pp.270-272.
87. D.A. Zindenbeck, Ind. Diamond Rev., **3**, 84-88 (1974).
88. French Patent No.1142688, MKI B 24 d 14/21. M.C. Soulard, Publ. 20.09.57.
89. West German Patent No. 936619, MKI B 24 d 17/00, C 04 b 31/16. A. Bergstrom, Publ. 15.09.55.
90. USA Patent No. 3585013, MKI B 24 d 3/10. D.K. Bruschek, Publ 15.06.71.
91. French Patent No.2047116, MKI B 24 d 17/00. De Beers Consolidated Mines Ltd., Publ. 16.04.71.
92. W.T. McFarlen, Plating, **57**, No.1, 46-50 (1970).
93. USA Patent No.3779727, MKI B 24 d 3/10. R.H. Signi and H.M. Cohen, Publ. 18.12.73.
94. Yu.I. Nikitin, E.M. Chistyakov and V.V. Pogorelyi, Sintet Almazy, No.1, 52-55 (1970).
95. USA Patent No. 2411867, MKI B24d 3/10, B. Brenner, Publ. 03.12.46.
96. French Patent No.1542036, MKI B 24 d 3/10. Ind. Diam. Lim., Publ. 11.10.68.
97. West German Patent No.1256070, MKI B 24 d 3/10 C 04 b 31/160. J.J. Holba and O. Hautzenthal, Publ. 07.12.67.
98. USA Patent No.3520667, MKI B 24 d 17/00. K.M. Taylor, Publ. 14.07.70.
99. USA Patent No.3525432, MKI B 07 d 5/344. R.F. Zurcher, Publ. 25.08.70.
100. P.P. Otopkov, A.I. Senchakov and V.M. Zubkov, Almazy, No.4, 1-3 (1969).

101. USA Patent No.3871840, MKI B 24 d 3/06. A.G. Wilder and H.C. Bridwell, Publ. 18.03.75.
102. West German Patent No.936619, MKI B 24 d 17/00, C 04 b 31/16. E.A. Bergstrom, Publ. 17.11.55.
103. Author's certificate No.414052 (USSR), MKI B 24d 3.06. A method of producing metallised diamonds, Otkrytiya. Izobreteniya (1974), No. 5, 40.
104. West German Patent No.2059371, MKI B24 d 3/10. J. Niel., Publ. 02.12.70.
105. Author's certificate No.496169 (USSR), MKI B 24 d 3/06. A method of producing aggregated diamond grains, Otkrytiya. Izobreteniya, (1975), No.47 (50).
106. B.M. Emel'yanov, Yu.N. Semenov and L.P. Kogosov, Sintet. Almazy, No.1, 55-58 (1970).
107. Yu.V. Naidich and I.A. Lavrinenko, Metallising and aggregation of diamond powders by adhesion-active metallic melts. In: Physical chemistry of surface phenomena, Naukova dumka, Kiev (1971), 220-223.
108. Yu.V. Naidich, I.A. Lavrinenko and Ya.F. Motsak, Aggregated metallised diamond powders in tools with an anti-friction organic binder. In: Synthetic diamonds - a key to technical progress, Institute of Superhard Materials, Academy of Sciences of the Ukrainian SSR (1974), pp.6-11.
109. Author's certificate No.569445 (USSR), MKI B 24 d 3/06. An alloy for coatings of abrasive materials (1977), No.31, 37.
110. West German Patent No.2021399, MKI C 04 b 31/16. R.A. Rose, Publ. 24.01.74.
111. French Patent No.2177031, MKI B 24 d 14/21. I. Dewar., Publ. 07.12.73.
112. West German Patent No.970535, MKI C 04 b 30/17, B 24 d 3/10. A. Ziden and O. Malmborg, Publ.24.11.41.
113. Kh.G. Tkhagapsoev, Z.J. Berov and A.A. Goov, Almazy i Sverkhtverdye Materialy, No.10, 10-12 (1971).
114. V.M. Lunev and V.P. Samoilov, Sintet. Almazy, No.4, 28-29 (1977).
115. Yu.V. Naidich, G.A. Kolesnichenko and B.D. Kostyuk, Poroshkovaya Metallurgiya, No.9, 91-94 (1973).
116. Yu.V. Naidich, B.D. Kostyuk, G.A. Kolesnichenko, et al, Poroshkovaya Metallurigya, No.12, 55-61 (1973).
117. English Patent No.614396, MKI B 23 K 19/00. T.W. Penrie, et al, Publ. 08.06.75.
118. USA Patent No.3351543, MKI B 24 d 3/10. A. Thomas and T.A. Vanderslice, Publ. 07.11.67.
119. West German Patent No.1502619, MKI B 24 d 3/10, B 24 d 17/00. E.H. Hull and T.A. Vanderslice, Publ. 22.05.69.
120. Ya.L. Potapenko, N.A. Paskal', A.A. Paskal' and V.N. Myslina, Sverkhtverdye Materialy, No.3, 17-20 (1992).
121. N.V. Pleshivtsev, Cathode sputtering, Atomizdat, Moscow (1968).
122. Yu.V. Naidich, V.P. Umanskii and I.A. Lavrinenko, Zavod. Lab. **43**, No.4, 504-505 (1977).
123. West German Patent No.2034522, MKI B 24 d 3/10. H.I. Wiemann, Publ. 20.01.72.
124. USA Patent No.3596649, MKI B 24 d 3/10. O.E. Olivieri, Publ. 03.08.71.
125. USA Patent No.3650714, MKI B 24 d 17/00. P. Farkas, Publ. 21.03.72.
126. Author's certificate No.246262 (USSR), MKI B 24 d 3/06. Otkrytiya Izobreteniya (1969), No.20, 42.
127. Yu.V. Naidich, I.A. Lavrinenko, Ya.F. Motsak, et al, Adgeziya Rasplavov

i Paika Materialov, No.2, 58-70 (1977).
128. Author's certificate No.507434 (USSR), MKI B 24 d 3/06.Otkrytiya Izobreteniya (1976), No.11, 58.
129. Author's certificate No.526678 (USSR), MKI B 24 d 3/06. Otkrytiya Izobreteniya (1976), No.32, 72.
130. N.V. Manukyan, A.P. Oganyan and G.S. Apoyan, Sverkhtverdye Materialy, No.5, 35-38 (1984).
131. A.P. Oganyan, Poroshkovaya Metallurigya, No.9, 33-37 (1991).
132. A.P. Oganyan, Poroshkovaya Metallurgiya, No.11, 11-13 (1991).
133. Yu.V. Naidich, V.P. Umanskii, I.A. Lavrinenko, et al, Almazy i Sverkhtverdye Materialy, No.3, 1-3 (1978).
134. V.G. Chuprina, V.P. Umanskii, I.A. Lavrinenko, et al, Poroshkovaya Metallurgiya, No.6, 79-83 (1980).
135. V.G. Chuprina, V.P. Umanskii and I.A. Lavrinenko, Ukr. Khim. Zhurn., **49**, No.4, 351-355 (1983).
136. Yu.V. Naidich, V.P. Umanskii, V.G. Chuprina, et al, Almazy i Sverkhtverdye Materialy, No.2, 2-4 (1979).
137. Yu.V. Naidich, V.P. Umanskii and I.A. Lavrinenko, Sverkhtverdye Materialy, No.6, 1-3 (1984).
138. V.G. Chuprina and V.V. Shurkhal, Poroshkovaya Metallurgiya, No.3, 96-100 (1987).
139. V.P. Umanskii, E.A. Zin'kova and V.Ya. Sabaldyr', Sverkhtverdye Materialy, No.6, 4-5 (1995).
140. V.P. Umanskii, V.L. Litvin and V.Ya. Sabaldyr', Adgeziya Rasplavov i Paika Materialov, No.30, 90-93 (1993).
141. Storms E., Refractory carbides, Atomizdat, Moscow (1970).
142. Index to the x-ray powder data film copyright by the American Society for Testing Materials, Philadelphia (1973), p.1000.
143. V.V. Shurkhal, V.P. Umanskii, I.A. Lavrinenko, et al, Adgeziya Rasplavov i Paika Materialov, No.23, 83-87 (1990).
144. B.A. Kolacheva and Yu.V. Levinskii (eds), Constants of interaction of metals with gases, a Handbook, Metallurgiya, Moscow (1987).
145. G.V. Samsonov (ed), Physico-chemical properties of oxides, a Handbook, Metallurgiya, Moscow (1969).
146. M.P. Arbusov, B.V. Khaenko, E.T. Kachkovskaya, et al, Ukr. Fiz. Zhurn., **19**, No.3, 497-501 (1974).
147. E.F. Ormont (ed), Variable composition compounds, a handbook, Khimiya, Leningrad (1969).
148. V.I. Kostikov, A.V. Nozhkina and Yu.A. Shesterin, Almazy, No.3, 1-3 (1971).
149. A.V. Nozhkina, Interaction of diamond with metals in sintering of powders, In: New directions in the development of diamond treatment, Scientific Research Institute of Diamonds, Moscow (1981), pp.10-19.
150. B.V. Khaenko and V.G. Fak, Izv. AN SSSR. Ser. Neorgan. Materialy, **14**, No.7, 1294-1301 (1978).
151. B.V. Khaenko, Izv. AN SSSR. Ser. Neorgan. Materialy, **15**, No.11, 1952-1960 (1979).
152. V.M. Lunev, V.P. Samoilov, V.P. Zubarev, et al, Sintet. Almazy, No.11, 26-28 (1978).
153. M.M. Ioffe, M.A. Maurakh and A.V. Nozhkina, Problem of interaction of diamonds with transition metals, In: Theory and practice of diamond treatment, Scientific Research Institute of Engineering, Moscow (1969),

pp.17-21.
154. N.V. Tsypin, E.S. Simkin and G.D. Kostenetskaya, Examination of phases formed at high temperatures as a result of interaction of diamonds with metals, In: Physico-chemical problems of synthesis of superhard materials, Institute of Superhard Material, Academy of Sciences of the Ukrainian SSR, Kiev (1978), pp144-147.
155. L. Sharpe adn H. Schonhorn, Adv. Chem. Ser., **43**, 189-201 (1964).
156. S.S. Voyutskii and B.V. Deryagin, Kolloid. Zhurn. **27**, No.4, 624-626 (1965).
157. B.V. Deryagin, N.A. Krotova and V.P. Smilga, Adhesion of solids, Nauka, Moscow (1973).
158. P.P. Kobeko and F.I. Marei, Zhurn. Tekhn. Fiziki, **16**, No.3, 277-282 (1946).
159. P.P. Kobeko, Amorphous substances, Publishing House of the Academy of Sciences of the USSR, Moscow-Leningrad (1952), p.432.
160. W. Weiss, Glastechn. Ber., **29**, No.10, 386-392 (1956).
161. J. O'Brien and G. Ryde, J. Amer. Ceram. Soc., **47**, No.1, 5-8 (1964).
162. P. Benjamin and C. Weaver, Proc. Roy. Soc. A., **254**, 177-181 (1960).
163. V.N. Batygin, A.M. Reshetnikov and I.M. Metelkin, Vacuum-tight ceramics and its bonding to metals, Energiya, Moscow (1973).
164. E. Krechmar, Spraying of metals, ceramics and plastics, Mashinostroenie, Moscow (1966).
165. A. Khasui, Spraying technology, Mashinostroenie, Moscow (1975).
166. V.G. Berezin, V.A. Martynov, V.P. Petrenko, et al, Zavod. Lab., **40**, No.2, 226-228 (1974).
167. Specification for adhesion tests of hot sprayed coatings, Weld. World, **11**, No.11, 291-294 (1973).
168. Z. Schweisstechn., **64**, No.6, 142-145 (1974).
169. B.M. Zakharov and M.G. Trofimov, Erosion-resistant coatings based on tungsten carbide and tungsten carbide with cobalt, In: Temperature resistant protective coatings, Nauka, Leningrad (1968).
170. I.E. Petrunin, I.Yu. Markova and A.S. Epatova, Metal science of brazing, Metallurgiya, Moscow (1976).
171. R. Williams and R. Backus, J. Appl. Phys., **20**, No.1, 98-106 (1949).
172. Author's certificate No.183459 (USSR), MKI B 24 d 3/06. Method of determining the strength of bonding of a coating with the parent metal, Otkrytiya. Izobreteniya, No.13, 44 (1966).
173. P.A. Pepchuk, Zavod. Lab., **30**, No.1, 93-94 (1964).
174. A.M. Pashennykh, Zhurn. Prikl. Khimii., **38**, No.5, 1181-1183 (1965).
175. E.I. Shilenkov, L.A. Shlapai and L.G. Shilenkova, Svaroch. Pr-vo, No.1, 40-41 (1973).
176. B.A. Lyashenko, V.V. Rishin, V.G. Zil'berberg, et al, Poroshkovaya Metallurigya, No.4, 96-100 (1969).
177. Yu.N. Polyakov, Zavod. Lab., **31**, No.10, 1249-1251 (1965).
178. N.G. Gladkov, A.M. Naumov, V.S. Sery, et al, Probl. Prochnosti., No.7, 120-122 (1976).
179. I.N. Katkov, L.A. Kuritsin and L.P. Orlov, Svaroch. Proiz., No.7, 40-41 (1973).
180. O. Heavens, J. Phys. Radium, **11**, No.7, 355-360 (1950).
181. C. Weaver and P. Benjamin, Proc. Roy. Soc., **77**, No.12, 163-176 (1960).
182. G. Gummenik (ed), High temperature inorganic coatings, Metallurgiya, Moscow (1968).

183. J. Beans, In: Technical Proceedings of the 43rd Annual Convention of the American Electroplaters Society, Washington (1956), pp.211-214.
184. S. Mosses and R. Witt, Amer. Chem. Soc., **41**, No.10, 2334-2338 (1949).
185. Yu.V. Naidich, A.M. Martirosyan and G.A. Kolesnichenko, Poroshkovaya Metallurgiya, No.4, 60-66 (1992).
186. Yu.V. Naidich, V.P. Umanskii, I.A. Lavrinenko, et al, Adgeziya Rasplavov i Paika Materialov, No.19, 73-77 (1987).
187. V.P. Umanskii, V.P. Brodskii, I.A. Lavrinenko and Yu.V. Naidich, Adgeziya Rasplavov i Paika Materialov, No.31, 93-96 (1994).
188. Yu.V. Naidich, V.P. Umanskii and I.A. Lavrinenko, Sverkhtverdye Mateialy, No.5, 18-22 (1991).
189. S.P. Yatsenko, Gallium. Interaction with metals, Nauka, Moscow (1974).
190. A.P. Semenov, V.V. Pozdnyakov and L.B. Kraposhina, Friction and contact interaction of graphite and diamond with metals and alloys, Nauka, Mosow (1974).
191. N.E. Charnaya, Almazy i Sverkhtverdye Materialy, No.7, 4-5 (1978).
192. A.A. Bugaev, Examination of metallised diamonds, in: Diamonds and hard alloy tools in mining, Tekhnika, Kiev (1965).
193. V.P. Umanskii, I.A. Lavrinenko, V.G. Chuprina and Yu.V. Naidich, Almazy i Sverkhtverdye Materialy, No.6, 1-2 (1982).
194. Yu.V. Naidich, V.P. Umanskii and I.A. Lavrinenko, Examination of the mechanical strength of bonding of metallic coatings and alloys on the surface of diamonds, In: Intl. Seminar: Superhard Materials (Kiev, June 1981), Institute of Superhard Materials, Academy of Sciences of the Ukrainian SSR, Vol.1, Kiev (1981), pp.217-218.
195. Yu.V. Naidich, V.P. Umanskii and I.A. Lavrinenko, Adgeziya Rasplavov i Paika Materialov, No.7, 82-86 (1981).
196. Yu.V. Naidich, V.V. Shurkhal, I.A. Lavrinenko and V.P. Umanskii, Adgeziya Rasplavov i Paika Materialov, No.14, 84-86 (1985).
197. A.N. Zaidel', Elementary estimates of measurement errors, Nauka, Leningrad (1968).
198. V.V. Nalimov, Using mathematical statistics in analysis of substances, Publishing House of the Physical and Mathematical Institute, Moscow (1960).
199. USA Patent No.392943, MKI B 24 d 17/00. R. Caveney, Publ. 30.12.75.
200. V.V. Shurkhal, I.A. Lavrinenko and V.P. Umanskii, Adgeziya Rasplavov i Paika Materialov, No.27, 74-79 (1992).
201. V.V. Shurkhal and I.A. Lavrinenko, Adgeziya Rasplavov i Paika Materialov, No.25, 79 (1991).
202. Properties of elements, a Handbook, in 2 volumes, Metallurgiya, Moscow (1976).
203. V.V. Shurkhal, V.P. Umanskii, I.A. Lavrinenko and E.I. Golovko, Adgeziya Rasplavov i Paika Materialov, No.26, 84-87 (1991).
204. V.V. Shurkhal, V.P. Umanskii, I.A. Lavrinenko and V.G. Chuprina, Adgeziya Rasplavov i Paika Materialov, No.18, 38-41 (1987).
205. M.S. Vinarskii and M.V. Lur'e, Experiment design in technological investigations, Tekhnika, Kiev (1975).
206. Yu.V. Naidich, V.P. Umanskii and I.A. Lavrinenko, Almazy i Sverkhtverdye Materialy, No.12, 1-4 (1980).
207. Yu.V. Naidich, et al, Ind. Diamond Rev., No.6, 327-331 (1984).
208. M.A. Krishtal and V.S. Goncharov, Zashchit. Pokrytiya na Metallakh, No.14, 22-26 (1980).

209. V.P. Umanskii, Yu.V. Naidich, I.A. Lavrinenko, et al, Almazy i Sverkhtverdye Materialy, No.5, 1-3 (1981).
210. F.A. Roal, Nature, **185**, No.4712, 523-524 (1960).
211. D.B. Fischbach, Chemistry and physics of carbon, P.L. Walker and P.A. Thrower (eds), **7**, No.4, 1-105 (1971).
212. V.R. Hower, Proc. Phys. Soc., **80**, No.3, 648-662 (1962).
213. A.A. Khomenko, L.T. Gankevich, B.G. Ostronov, et al, Sintet. Almazy, No.1, 9-14 (1974).
214. J.W. Harris and E.R. Vance, Contr. Miner. and Petrol, **35**, No.3, 227-234 (1972).
215. D.V. Fedoseev, V.L. Bukhovets, S.P. Vnukov and B.A. Anikin, Graphitisation of diamond at high temperatures, In: Surface and thermophysical properties of diamonds, Institute of Superhard Materials, Academy of Sciences of the Ukrainian SSR, Kiev (1985), pp.6-9.
216. D.V. Fedoseev, S.P. Vnukov, V.L. Bukhovets, et al, Dokl. AN SSSR, **27**, No.1, 149 (1984).
217. V.G. Malogolovets and E.G. Gatilova, Sintet. Almazy., No.4, 29-32 (1970).
218. P.P. Otopkov and A.V. Nozhkina, Effect of heat treatment on the physico-chemical properties of diamonds, In: physico-chemical properties of diamonds, Scientific Research Institute of Engineering, Moscow (1974), 15-32 (Tr. VNIIA Almaza; No.3).
219. K.J. Rone, et al, Amer. Ceram., **50**, No.9, 791-792 (1971).
220. A.V. Belyankina, L.M. Mul'takh, E.S. Simkin, et al, Sintet. Almazy, No.1, 20-22 (1972).
221. N.V. Tsypin and S.A. Bozhko, Sverkhtverdye Materialy, No.1, 15-18 (1982).
222. G.A. Meerson, E.G. Zimina and V.V. Zhuravlev, Examination of oxidation of diamond powders, In: Diamonds and diamond tools, Almazy i Almaz. Instrument., Moscow, 17-23 (1973).
223. J. Thomas, Examination of the oxidation of graphite by microscopy, In: Chemical and physical properties of carbon, Mir, Moscow (1969).
224. M.F. Gukasyan and E.A. Konorova, Almazy, No.1, 1-6 (1973).
225. F. Walker (ed), Chemical and physical properties of carbon, Mir, Moscow (1969).
226. N.I. Kornilov, V.V. Pimenov and N.D. Suslikova, Almazy i Sverkhtverdye Materialy, No.10, 3-4 (1978).
227. B. Yoyng, Ind. Diamond Rev., No.11, 483-488 (1966).
228. V.M. Titova and S.I. Futergendler, Kristallografiya, **7**, No.6, 926-929 (1962).
229. Yu.F. Shul'pyakov and Yu.S. Genshaft, Interaction of diamonds with different substances at high pressures and temperatures, In: Physico-chemical properties of diamonds, (Tr. VNIIA Almaza), Scientific Research Institute of Engineering, Moscow (1974), 50-58.
230. R.M. Horton and M.D. Horton, High temp.- high pressure, **4**, No.1, 39-48 (1972).
231. T.N. Loladze and G.V. Bokuchava, Wear of diamonds and diamond wheels, Mashinostroenie, Moscow (1967).
232. T.N. Loladze, Tr. Tbil. Politekhn. in-ta No. 7, 109-114 (1973).
233. C.A. Brookes, Ind. Diamond Rev., No.1, 21-24 (1971).
234. T. Tanaka, Bull. Jap. Soc. Precis. Eng., **14**, No.2, 107-108 (1980).
235. English Patent No.1421823, MKI B 24 d 5/00. H. Matthews, Publ. 27.01.76.

236. A.S. Vishnevskii, A.V. Lysenko, T.D. Ositinskaya and V.G. Delevi, Izv. AN SSSR. Neorgan. Materialy, **11**, No.1, 1589-1593 (1975).
237. A.S. Vishnevskii and A.V. lysenko, Sintet. Almazy, No.4, 7-9 (1974).
238. Yu.D. Klebanov and V.N. Sumarokov, Fiz. Khim. Obrab. Mater., No.3, 150-151 (1975).
239. A.E. Gorodetskii, L.L. Builov, V.M. Luk'yanovich, et al, Fiz.-khim. Probl. Kristallizatsii, No.2, 62-67 (1971).
240. N.V. Tsypin, E.S. Simkin, L.M. Mul'takh, et al, Sintet. Almazy, No.5, 33-36 (1971).
241. E.S. Simkin, Adgeziya Rasplavov i Paika Materialov, No.16, 90-94 (1986).
242. V.B. Fedorov, M.Kh. Shorshorov and D.K. Khakimova, Carbon and its interaction with metals, Metallurgiya, Moscow (1978).
243. R. Cahn (ed), Physical metallurgy, Mir, Moscow (1968).
244. D.V. Fedoseev, Kolloid. Zhurn., **40**, No.2, 402-403 (1978).
245. V.N. L'nyanoi and S.A. Chepur, Izv. Vuzov. Cher. Metallurgiya, No.7, 94-98 (1980).
246. A.M. Kuz'min, L.M. Terent'eva, T.D. Firsova, et al, Fizika i Khimiya Obrab. Materialov., No.6, 136-139 (1977).
247. V.E. Ovcharenko, L.A. Turovets, E.F. Dudarev, et al, Poroshk. Metallurgiya, No.4, 57-62 (1980).
248. Author's Certificate No.371062, MKI B 24 3/06, Mul'takh L.M, et al, Otkrytiya Izobreteniya, No. 12, 37 (1973).
249. R. Sappok and H.P. Bohm, Carbon, No.6, 283-295, 573-588 (1968).
250. Yu.V. Naidich, V.V. Shurkhal, V.P. Umanskii and I.A. Lavrinenko, Dokl. AN Ukrainy. Ser. A. Fizmat. i Tekhn. Nauki., No.2, 72-76 (1990).
251. S.Kh. Lifshits and A.P. Grigor'ev, Mechanism of catalytic hydration of diamond in the presence of nickel, In: Examination in the area of interaction of different forms of carbon with gas or liquid medium, Institute of Superhard Materials, Academy of Sciences of the Ukrainian SSR, pp.8-16 (1986).
252. A.P. Grigor'ev, S.Kh. Lifshits and P.P. Shamaev, Kinetika i Kataliz, **18**, No.4, 948-952 (1977).
253. A. Tomita, N. Sato and I. Tamai, Karbon, **12**, No.2, 143-149 (1974).
254. A.R. Andrievskii and I.I. Spivak, Strength of refractory compounds and materials based on them, a Handbook, Metallurgiya, Chelyabinsk (1989).
255. M.E. Drits (ed) Properties of elements, a Handbook, Metallurgiya, Moscow (1985).
256. V.M. Krivoruchko, V.V. Sagalovich and N.S. Poltavtsev, Stresses in molybdenum coatings produced by thermal dissociation of $Mo(CO)_6$, In: Protective high temperature coatings, Nauka, Moscow (1972).
257. G.I. Zhuravlev, Poroshkovaya Metallurgiya, No.10, 36-43 (1966).
258. S.B. Aimbinder and E.F. Klokova, Dokl. AN SSSR, **146**, No.5, 1058-1060 (1962).
259. A.A. Appen, Main physico-chemical principles of development of creep resisting inorganic coatings, In: Heat resisting coatings, Zharostoikie Pokrytiya, Leningrad and Nauka, Moscow (1965).
260. V.A. Podoroga, V.P. Kebko, V.S. Morganyuk, et al, Probl. Prochnosti, No.4, 88-93 (1990).
261. V.A. Podoroga, V.P. Kebko, V.S. Morganyuk, et al, Probl. Prochnosti, No.4, 93-97 (1990).
262. Nishikawa T. and Kawakita T., Bul. Jap. Soc. Precis. Eng., **20**, No.2, 125-126 (1986).

263. Ind. Diam. Rev., **52**, No.551, A9 (1992).
264. K. Steinmetz, et al, Ind. Diam. Rev., **49**, No.533, 164-167 (1989).
265. D.J. Dansten, Pribory dlya Nauch. Issled., No.6, 165-167 (1991).
266. M. Seal, Mater. Sci. and Eng. B., **11**, No.1/4, 167-171 (1992).
267. GOST 19738-74. Silver brazing alloys, Submitted 01.01.75.
268. Author's certificate No.634886, MKI3 B 23 K 1/20. A brazing alloy for brazing jewelry components, Otkrytiya. Izobreteniya, No.44, 24 (1978).
269. Japanese Patent No.54-42868, MKI B 23 K 35/30, C 22 C 9/04. A method of fixing diamond grains on metallic base, T. Noloji, Published 17.12.79.
270. Japanese Patent No.55-19719, MKI B 23 K, 35/28, C 15/00. A brazing alloy. Sumitamio Kindzo Kosan K K, Published 28.05.80.
271. Author's certificate No.650759 (USSR), MKI B 23 K 15/20. A brazing alloy for brazing cermet diamond-bearing matrices with the steel casing of the tool, Otkrytiya. Izobreteniya, No.9, 41 (1979).
272. N.I. Kornilov and A.D. Kotok, Razvedka i Okhrana Nedr. No.5, 25-27 (1980).
273. N.F. Lashko and S.V. Lashko, Brazing of metals, Mashinostroenie, Moscow (1977).
274. French Patent No.1073956, MKI B 24 d 3/10. N.V. Philips, Publ. 27.05.54.
275. French Patent No.1240869, MKI B 24 d 14/21. N.V. Philips, Publ. 09.09.60.
276. French Patent No.1240395, MKI B 24 d 17/00. N.V. Philips, Publ. 02.09.60.
277. English Patent No.932729, MKI B 28 C, B 23 K. T. Threadgold, Publ. 31.07.63.
278. Yu.V. Naidich, Yu.N. Chuvashov and V.S. Zhuravlev, Zhurn. Fiz. Khimii, **52**, No.10, 2694-2696 (1978).
279. English Patent No.1310963, MKI B 23 K 19/00. C. Clark, Publ. 21.03.73.
280. French Patent No.2240791, MKI B 24 d 17/00. T. Lowder and E. Tansch, Publ. 29.03.74.
281. Author's certificate No.567574 (USSR), MKI B 23 K 1/19. A brazing alloy for brazing components made of high hardness materials, Otkrytiya. Izobreteniya, No.29, 42 (1977).
282. Author's certificate No.565070 (USSR), MKI C 22C 19/08. A nickel-based alloy, Otkrytiya. Izobreteniya, No. 26, 55 (1977).
283. USA Patent No.4448605, MKI3 C 22C 30/00. M. Howard, Publ. 02.12.82.
284. English Patent No.1423238, MKI B 23 K 19/00. F. Redfern, Publ. 04.02.76.
285. Author's certificate No.536905 (USSR), MKI3 B 23 K 1/20. A method of brazing hard alloy materials, Otkrytiya. Izobreteniya, No.44, 35-36 (1976).
286. G. Manko, Brazing and brazing alloys, Mashinostroenie, Moscow (1968).
287. I.E. Petrunin (ed), A handbook of brazing, Mashinostroenie, Moscow (1984).
288. Prices of industrial furnaces and heating systems, Preiskurantizdat, Moscow p.144 (1981).
289. USA Patent No.3868750, MKI B 24 d 3/06, 3/10. G. Ellis, Publ. 04.03.75.
290. V.P. Umanskii, V.V. Shurkhal, I.A. Lavrinenko and Yu.V. Naidich, Almazy, No.1, 8-12 (1984).
291. I.A. Lavrinenko, Yu.V. Naidich, Ya.F. Motsak, et al, Almazy i Sverkhtverdye Materialy, No.9, 6-9 (1976).
292. I.A. Knyazeva, M.S. Pivovarov, V.P. Baraban, et al, Tools based on dia-

monds with metallic high hardness coatings for processing of glass. In: New developments in the theory and practice of production and application of synthetic superhard materials in the national economy. Institute of Superhard Materials, Academy of Sciences of the Ukrainian SSR, Vol.3, Kiev (1978).

293. G.A. Kolesnichenko, E.M. Chistyakov and Ya.F. Motsak, In: Physical chemistry of condensed phases of superhard materials and their interfaces, Naukova dumka, Kiev (1975).

294. USA Patent No.2216828, MKI B 24 d 3/10. T. Kurotomi, Publ. 02.11.71.

295. English Patent No.12477664, MKI B 24 d 3/10. C. Clark, Publ. 10.09.71.

296. West German Patent No.1243097, MKI3 B 24 d 3/10. E. Baas, Publ. 15.09.65.

297. Ya.F. Motsak, I.A. Lavrinenko, S.M. Pivovarov, et al, Almazy i Sverkhtverdye Materialy, No.10, 7-10 (1975).

298. Yu.V. Naidich, G.P. Volk and I.A. Lavrinenko, Poroshkovaya Metallurgiya, No.9, 22-25 (1981).

299. Yu.V. Naidich, G.P. Volk, L.Yu. Ostrovskaya, et al, Poroshkovaya Metallurgiya, No.6, 79-82 (1988).

300. I.A. Knyazeva, V.P. Baraban and I.A. Lavrinenko, Adgeziya Rasplavov i Paika Materialov, No.10, 84-87 (1982).

301. Yu.V. Naidich, I.A. Lavrinenko, G.P. Volk and V.P. Umanskii, Examination and development of adhesion-active binders for diamond tools produced by powder metallurgy methods, In: Proc. XIIth Nat. Powder Metallurgy Conf., Riga, Lat. INTI 27-31 (1975).

302. Yu.V. Naidich, I.A. Lavrinenko, S.B. Zadoyan, et al, Metallising of diamonds and adhesion-active binders for diamond tools, In: Physical chemistry of condensed phases, superhard materials and their interfaces, Naukova dumka (1975).

303. USA Patent No.5062865, MKI3 B 24 d 3/00. U. Chen and Ch.M. Sung, Publ. 05.11.91.

304. S.A. Ashinov, Z.Zh. Berov, A.G. Martynov, et al, Sverkhtverdye Materialy, No.1, 36-40 (1993).

305. A.T. Collins, et al, Ind. Diam. Rev., No.350, 19-22 (1970).

306. V.F. Ivanov, Yu.A. Detchuev and A.A. Fotchenkov, Adgeziya Rasplavov i Paika Materialov, No.4, 92-93 (1979).

307. Yu.V. Naidich, V.M. Chashnik, I.A. Lavrinenko, et al, Poroshkovaya Metallurgiya, No.8, 52-56 (1987).

308. V.M. Chashnik, I.A. Lavrinenko, G.P. Volk, et al, Mechanism of formation and protective properties of chromium and chromium-nickel coatings on diamond, In: Structural chromium alloys. Tekhnologiya, Naukova dumka, Kiev (1988).

309. Z.J.Berov, B.S. Karamursov, A.Kh. Tlibekov, et al, Sverkhtverdye Materialy, No.5, 55-61 (1998).

310. N.A. Bondarenko, N.V Novikov, V.A. Mechnik, et al, Sverkhtverdye Materialy, No. 6, 3-15 (2004).

311. V.S. Antonyuk, B.A. Lyashenko, E.B. Soroka, et al, Sverkhtverdye Materialy, No. 2, 72-75 (2005).